Second Edition

Wetland Landscape Characterization

Practical Tools, Methods, and Approaches for Landscape Ecology

Second Edition

Wetland Landscape Characterization

Practical Tools, Methods, and Approaches for Landscape Ecology

Ricardo D. Lopez
John G. Lyon
Lynn K. Lyon
Debra K. Lopez

CRC Press
Taylor & Francis Group
Boca Raton London New York

CRC Press is an imprint of the
Taylor & Francis Group, an **informa** business

CRC Press
Taylor & Francis Group
6000 Broken Sound Parkway NW, Suite 300
Boca Raton, FL 33487-2742

First issued in paperback 2017

© 2013 by Taylor & Francis Group, LLC
CRC Press is an imprint of Taylor & Francis Group, an Informa business

No claim to original U.S. Government works

ISBN-13: 978-1-4665-0376-2 (hbk)
ISBN-13: 978-1-138-07609-9 (pbk)

Visit the Taylor & Francis Web site at
http://www.taylorandfrancis.com

and the CRC Press Web site at
http://www.crcpress.com

Contents

Preface

Decision makers of all types are increasingly called upon to determine the details of the complex and transitional ecosystem called "wetlands." Because wetlands are, by their very nature, ephemeral and transitional, their complexity makes the task of capturing their essence, that is, characterization, very challenging. The need for characterizing wetlands is growing every day to meet scientific and societal needs, and this need is on the increase as we better understand the wealth of ecosystem services wetlands provide us. For those continuing their quest for wetland knowledge, this second edition of *Wetland Landscape Characterization* is designed to enhance their knowledge base, providing them with a pathway to understanding how wetland characterization tools, methods, and approaches can be integrated to address twenty-first century wetland issues.

Authors

Ricardo "Ric" Daniel Lopez is a leader in the field of wetlands ecology and landscape ecology. During his tenure in academia and public service, he has lead in geographically diverse applications of field-based and geospatial approaches to theoretical and applied environmental topics. This body of work includes monitoring and assessing of wetlands and streams, invasive plant species, broad-scale indicators of sustainability, and solutions to risk-based issues. A native of coastal California, Lopez spent his youth knee-deep (or deeper) in the many wetlands and tide pools of the region. Lopez earned his BS in ecology, behavior, and evolution at the University of California, San Diego, and his master's and doctoral degrees in environmental sciences at The Ohio State University, with an emphasis on wetland ecology and landscape ecology.

John Grimson Lyon was interested early on in wetlands as places of native vegetation. This interest was honed during youthful wanderings in the mountains and river valleys of the Pacific Northwest, California, Nevada, and Alaska. Systematic study at the undergraduate, graduate, and professional levels has yielded a body of work on remote sensing, mapping, identification, and characterization of processes in wetlands, and related ecosystems in the Great Lake states and western United States.

Lynn Krise Lyon is a lifelong educator, writer, and artist. She spent a good portion of her youth playing in creeks and streams. As an adult, she has visited wetlands all over the United States with her husband, John. She abhors black flies and snakes, loves cranberry bogs, and fervently believes Michigan has the best wetlands in the world.

Debra Kim Lopez has dedicated her existence to the appreciation of literature and writing, global sustainability issues, and the social sciences. She values the importance of global initiatives for improving communities around the world; she has traveled extensively. Lopez has partnered with her husband, Ric, on a plethora of wetland and other environmental issues, ever since they first encountered one another on a common travel adventure twenty-two years ago.

1

Introduction: The Challenge

One ponders the imponderable. What are the future challenges for landscape characterization particularly related to wetlands? You may have encountered these challenges in your professional life and observed them in your personal life too. Many of these challenges are related to the big issues in the news and are on the public's mind regarding adequate and safe water, sufficient and affordable food supplies, and infrastructural development. All have something to do with landscapes and wetlands.

Wetland landscape characterization is an important component of determining the degree to which wetlands improve environmental conditions in a particular location. Similarly, the type of wetland, the characteristics of particular types of wetlands, and a wetland's position in the landscape has a tremendous influence on water-quality benefits of aquatic resources. In essence, wetlands provide critical linkages between water and land, people and creatures, and their food and water as well as shelter. It is important to understand these critical linkages, and functions and services that wetlands provide.

The challenges and dialogue specifically focus on critical needs and linkages that are inextricably part of wetlands. The safety and security of water, which is a matter of survival, have linkages both on the issues of *water quality* and *water quantity*. Specifically, water sustains life and without it humans perish in a few days. Due to water's unique properties, water can convey pollutants directly to organs and cells. However, wetlands can store water and delay or reduce the flow of drainage or subsurface water flow from developed areas, consequently removing or reducing the amount and form of pollutants to improve water quality (i.e., chemistry), and thus improving water quality of streams, lakes, and seas. These linkages and societal benefits are intertwined, yet must be understood for thoughtful decision making.

Another critical linkage between society and wetlands is food safety and food security, particularly involving crop and grazing productivity, the availability of wild foods, and the security of the public's food supply (Thenkabail et al. 2009). Food supply and the security of the food supply is critical because without minimum caloric need being met, humans die in a matter of weeks, and reduced nutritional intake is a serious threat to individual and community health. Also because food is another primary conveyance of chemical constituents into organs, any chemical, biological, viral, or radiological contamination of food is a direct threat to individual and community health and longevity. Wetlands are therefore a critical link to food supplies and security, directly providing food; including hunted animals (e.g., goose and duck), wild and cropped foods (e.g., rice, fish, crustacean, and seaweeds) and pollination, and wetlands also directly improve water quality of source water to cropped areas.

A third critical linkage between society and wetlands is generally described as *infrastructure,* which is to say the interconnected framework that supports the development of humans' everyday physical environment, from personal dwellings, roads, and bridges to the Internet and other global communications networks. An emerging way of understanding this infrastructural linkage is by way of the concepts of ecological goods and services. These can include water supply regulation and filtration; protection from flood waters and storm surges; habitat provisioning for terrestrial and aquatic species; and recreational, spiritual, and aesthetic benefits to society, supporting the structure of communities and cultures.

Of course this infrastructural support varies tremendously throughout the world, both societal and geographic, depending on the traditions and values of people. In addition, local, regional, and global policies and decisions that involve wetlands or allied natural resources require social scientists, along with other scientists, economists, and others to provide decision makers with current and accurate information about wetland conditions and the risks to wetlands (that could lead to wetland degradation or outright loss).

This edition provides the methods and approaches necessary for informed decision making on wetland resource issues of all types. It provides several new examples and case studies that describe detailed and general lessons for all applications. This also integrates well with an emerging infrastructural, ecosystem goods and services perspective to better assist readers who may encounter these concepts and challenges during the assessment and characterization of wetlands, within the context of the larger landscape. And, importantly, supplies pithy scientific analyses to drive the dialogue.

Accordingly, this book increases the knowledge, skills, and abilities of professionals and apprentices, and provides for continued growth in their careers. It brings the reader the concepts and skills necessary to achieve a better understanding of how project goals can be best achieved in the rapidly changing disciplines of landscape sciences and wetland ecology and management, by providing explicit examples that illustrate a variety of encountered situations and solutions. Each of these examples offers more depth and breadth of information, particularly in terms of utilizing current techniques in assessment, inventory, and monitoring of natural resources under conditions that are ever-changing over space and time.

The discerning reader will find many opportunities to combine the conceptual and operational linkages described here, which outlines the integration of wetland landscape characterization, inventorying, monitoring, assessment, and restoration techniques. The inventive reader will also find several opportunities to make connections between their specific natural resource management issues and goals, and the specific examples, techniques, and approaches provided in subsequent chapters.

2

Key Themes Driving Landscape Characterization

Wetlands are all about linkages. As a mixture of phenomena juxtaposed between terrestrial and aquatic landscapes, they have challenged methods and approaches for their analyses. To inform the public and decision makers it is necessary to use technologies to probe the landscape and linkages. Any number of questions can be addressed using these approaches with the wide-ranging variety of tools, techniques, and approaches available to the practitioner.

Wetland Principles

Wetlands are land areas that are periodically flooded or covered with water (Figure 2.1). It is the presence of water at or near the soil surface for more than a few weeks during the growing season that may help to create many wetland conditions. The water slows diffusion of oxygen into the soil and to plant roots. Lack of oxygen or anaerobic conditions causes major changes in the soil chemistry. Only certain *wetland plants* are adapted to live in these harsh conditions. Their adaptations allow them to use available soil nutrients, and they exhibit a variety of physical and physiological adaptations to grow in the absence of available oxygen.

The combination of anaerobic and waterlogged soils, the presence of wetland plants, low-lying topography, and other conditions help to create a different land cover type called wetlands. These characteristics and conditions are also used to define and identify wetlands.

Different types of wetlands have been created by hydrological and topographical conditions. This has a lot to do with the variety of water bodies or sources of water associated with the wetland. For example, wetlands adjacent to rivers take on the characteristics of the riparian and riverine conditions. Wetlands on the shore of lakes have many hydrological characteristics that are driven by the lake system. Wetland areas on marine coasts have coastal characteristics and are also influenced by the varying salinity concentrations from open ocean, coastal ocean, and neighboring estuarine waters. Hence,

FIGURE 2.1
The combination of water and vegetation and standing water over soils make a wetland, such as this coastal wetland area of northern Lake Michigan.

the hydrology of a given area is important to the characteristics, conditions, the functions of wetlands, and ultimately to their identification.

As another example, the variability of temporary or ephemeral wetlands makes them particularly hard to distinguish. These areas may only display wetland function for a little over two weeks during the growing season. Yet they are vital to the ecology of most ecosystems, such as the desert or semidesert ecosystems in the western United States, Canada, and Mexico.

One central priority of the wetland issue is the definition of a wetland. In particular, we are interested in the definition of a jurisdictional wetland (Lyon 1993; Lyon and Lyon 2011). A wetland is defined in the U.S. Army Corps of Engineers (USACE) 1987 wetland manual as "areas that are inundated or saturated by surface or ground water at a frequency and duration sufficient to support, and that under normal circumstances do support, a prevalence of vegetation typically adapted for life in saturated soil conditions. Wetlands generally include swamps, marshes, bogs and similar areas."

A jurisdictional wetland is defined in the field. It must exhibit a predominance of wetland plants, soils subject to waterlogging or hydric soils, and indicators of wetland hydrology. These three individual criteria must be addressed, and areas exhibiting all three indicators are deemed jurisdictional wetlands (Lyon 1993; Lyon and Lyon 2011).

Trends in Wetland Inventory, Monitoring, and Assessment

Wetland science and landscape ecology has undergone tremendous change in the past decade, and the melding of the two disciplines has yielded new and exciting trends in inventory, monitoring, and assessment. New issues, new results, the advance of techniques, and decisive applications in wetland sciences and landscape ecology all have a bright future because of their increased applicability to fundamental societal needs, in terms of assessment and associated decision making. Simultaneously with this increased need for compelling information about wetlands and their status or conditions, the last ten years has witnessed important new developments and refinements of earlier paradigms, enhancing areas or landscapes, and the combination of biological, physical, and chemical processes (Figure 2.2). The emergent paradigm of the landscape sciences encompasses the disciplines and subdisciplines necessary to address the characteristics and ecology of wetlands at a landscape scale. Landscape science incorporates both the concepts of landscape ecology and landscape characterization.

Landscape characterization is the combination of methods and approaches that allows the characterization of earth and water resources, and the processes that drive the systems. Landscape ecology is the mix of biological, chemical, and physical processes, and characterizes the earth and water at the landscape scale (Figure 2.3). Together, wetland landscape characterization

FIGURE 2.2
Movement of water, such as this tidal flow, causes areas to exhibit partial wetland characteristics during the day and perhaps less clear characteristics at other times in coastal Oregon.

FIGURE 2.3
Ebb and flow of water on the surface of soils across a broad landscape often is enough to keep soils saturated beneath the surface in the absence of standing water in some areas of the back barrier marshes and wet soils of Galveston Island, Texas.

and wetland landscape ecology allows for the integration of theoretical and practical characterization, evaluation and prediction of landscape resources, and the dynamics of their processes as they relate to wetland ecosystems. Importantly, the relatively new *landscape approach* makes possible the incorporation of all processes, both sociological and ecological, for improved wetland analyses and decision making. The landscape sciences represent the culmination of years of study and theoretical development and practical geospatial technologies. The result is a set of concepts that can be used to formulate the problems, execute their study in nature, help in the parameterization and simulation of real processes in a mathematical manner, and the prediction of future trends and the risk that is posed by natural and anthropogenic stresses.

An approach that is currently used in the landscape sciences is that of environmental risk assessment. Risk assessment involves the characterization of specific risks and how those risks may influence or expose natural resources, such as a wetland, or receptors of the risk, such as organisms within a wetland. The exposure of a wetland (resource) or organism (receptor) to a natural or human-induced stress may often define the problem that is necessary (or desired) to address through decisions and actions. Risk can be determined by the characterization of exposures and ecological effects to receptors, by stressors.

Once determined, it becomes important to evaluate the concern through risk characterization and address the correction of the problems through

FIGURE 2.4
Many wetland areas appear problematic as compared to most people's ideal (such as in Figure 2.3). Areas with minimal plants but with anaerobic soils fit the definition of a wetland such as these areas on the coastal shore of Alaska.

risk management. To evaluate the potential risk or conduct risk assessment involves the determination of endpoints or assessment criteria. Endpoints come in two forms: either an assessment endpoint or a measureable endpoint. Often, endpoints of interest are difficult to evaluate, so that surrogate variables or measurable endpoints are used. Numerous examples of endpoints are provided in subsequent chapters.

Two important considerations in utilizing the risk assessment approach in landscape evaluations are spatial and temporal characteristics of the problem (Figure 2.4). Ecosystems vary over space and time, so the risk assessment paradigm should incorporate both of these dimensions and variables, particularly when utilizing *geospatial technologies* such as remote sensor data and geographic information systems (GIS).

Wetland landscape characterization (Figure 2.5) and wetland landscape ecology need to both encompass the risk assessment and characterization concepts necessary to measure, model, and predict the current, historical, and future status of wetlands and related ecosystems. Within this edition the concepts necessary to characterize (Figure 2.6) and evaluate the wetlands and related habitats, with examples, are demonstrated, including the methods and interpretations necessary to conduct similar efforts.

FIGURE 2.5
Presence of obligate wetland plants such as cattail often help identify the presence of wetlands even in a winter scene populated with plant residue in Ohio.

FIGURE 2.6
Wetlands and ponds supply recreation and learning opportunities that many people take advantage of including summer research programs in Michigan.

Wetland–Landscape Paradigm

In the United States and abroad there is burgeoning interest in maintaining the wetlands that are currently present in the landscape through a variety of means. This would include creating them in areas not previously found as well as restoring them in areas where they were previously found. This expansive interest is in addition to calls from public interest groups and regulations from resource management agencies. Numerous national and international consortia and interests also have influenced the burgeoning emphases on wetland protection and restoration. The powerful drivers of public need and governmental oversight and regulation make for a bright future for wetland analyses as well as management programs and projects.

Wetlands are very unique, and in some cases extremely rare, in that they are an unusual transitional ecological zone, sharing and mixing ecological components and functions from purely aquatic and purely terrestrial habitats. What one observes is often a mixture, or transitional gradient, of these upland or terrestrial conditions and aquatic conditions, which may be an added difficulty for interpretation due to the complex interspersion of soils, hydrologic conditions, and flora. Because of this mixture of terrestrial and aquatic characteristics, wetlands provide many functions and mixtures of functions of both aquatic and terrestrial ecosystems, creating the unique and rare environment that makes wetlands so special and treasured for their functional characteristics.

Most common or large wetland areas can be recognized by many people, however, the transitional character of wetlands and their potential impact on the larger watershed, or landscape, can be difficult to identify and to quantify. This sort of information is necessary for those who need to characterize wetlands and their beneficial impact on the larger landscape. This information quest is further complicated by the sometimes fleeting nature of wetlands, such as where wetlands are only temporary features in the landscape (e.g., only flooded part of the year), are directly adjacent to terrestrial ecosystems (e.g., an upland forest), or are directly adjacent to aquatic ecosystems (e.g., a stream) (Figure 2.7).

Although some people have a high comfort level about their knowledge of wetlands and believe they know how a given type of wetland should appear (Figure 2.8), others may need to utilize additional information to secure their decision about the existence of a wetland. A major concern is the lack of adequate understanding of wetland characteristics and functions in the face of a need to make land management decisions (Figure 2.9). A poor understanding of all the different wetland characteristics and functions can lead to nonsustainable land management decisions in a particular locale.

An important distinction is that landowners often identify wetland areas in a different manner than do regulators or other wetland delineation experts. Such regulators or experts are experienced at identifying less-than-apparent

FIGURE 2.7
Many areas experience floods or periods of standing water and saturated soils, and may have wetland functions for short periods of time, such as these rain-flooded areas of a farm field in South Dakota.

FIGURE 2.8
The riparian areas of the stream channel are often just that, but wetland plants abound and wetland soils are patchy in distribution dictated by historic meandering of water in interior Alaska.

FIGURE 2.9
These typical wetlands are understood by the layperson and include wetland-loving moose, illustrating the linkages of wetland landscapes to wetland functions and ecological services.

wetland characteristics and types. Wherever land areas have the potential to be waterlogged or flooded, there is potential for wetland presence (Figure 2.10). Depending on local and regional interests, activities in and around these potential wetlands may create harm and therefore may come under additional scrutiny.

The widespread distribution of wetlands holds the promise of the resource being found on many lands, including those lands that are slated for change in land cover or new management practices. Where such ecosystem risks may occur, personnel are working and naturally they will encounter wetlands in the course of practicing their profession. Where development and wetlands coexist, there are often potential or actual risks of loss of the physical wetland or loss of wetland functions (i.e., degradation of wetland processes). These losses are addressed by federal, state, and local regulations and oversight. Hence, interested parties must pay close attention to the potential for risks to wetlands, and it is extremely critical to know the location and variety of wetland ecosystems found in the landscape.

There are many opportunities to work with wetland ecosystems in such a way as to avoid or minimize disturbances, and make the improvement in the numbers and functions of wetlands to develop a more integrated and coordinated plan for land and water management using the landscape approach. The distributed nature of wetlands requires that their locations be identified for thoughtful management, preservation, and evaluation of risks. There are several ways to locate and inventory wetlands, and assess the risks imposed by stressors. It is necessary to accomplish these goals related to wetland

FIGURE 2.10
Spring runoff of snowmelt creates many wetland areas that may or may not be present later in the summer. Use of wetland indicators such as wetland plant residue can help locate resources across the landscape. Snow partially covering wetland vegetation is depicted in this image.

inventory and to do so in a manner that matches the needs of a particular community, paying particular attention to the financial resources human resources and skills that are available. There are also several different levels of details that can be collected about wetland areas to fully evaluate risks associated with human activities. These activities are all part of wetland landscape characterization and risk assessment.

Wetland landscape characterizations are particularly needed, and indeed required, for a number of reasons. Water quality and water quantity are naturally a unifying concern, with wetlands at the heart of the amalgam of physical hydrologic and chemical processes involved in these two important aspects. The mix of physiochemical processes, which are often mediated by wetland biogeochemistry, yields an amazing variety of functional properties that all life forms depend upon. To bring together the processes, functions, and uses of wetlands necessitates approaches and methods that are inherently spatial, and that incorporate concepts of environmental risk and environmental health.

The desire to better understand and manage wetland landscapes may create excellent opportunities to examine current capabilities for measurements, inventory, and modeling. Subjective judgments and anecdotal information no longer suffice to address the pressing issues of wetland characterization and associated decision making. In their place are new tools and approaches for providing more quantitative data and rigorous interpretations, including both conceptual and technological approaches.

FIGURE 2.11
The use of a variety of images, sensor types, and time of the year of seasons facilitates interpretation and leads to a convergence of evidence to support a hypothesis. Winter period images or low sun angle images can reveal additional information. Note the ice covering lakes helps to distinguish them from the surrounding land covers and displays their surface area. Dead plants and plant residue can be distinguished from forested areas by pattern and texture as well as gray tone.

We can now measure wetland and related ecosystems with a tremendous number of sample locations and at vast scales, relying on computing power to process and store this wealth of empirical data (Figure 2.11). Value-added processing of these data with simple or complex algorithms, based on theory and practice, allows for a quantitative and accurate view of a much larger extent of the landscape, with improved accuracy and precision as well. Empirical and simulation modeled results tested by assessments of accuracy now pace the dialogue in wetland landscape characterization. The power of these wetland-related decisions, along with facts and predictions behind them, is evidenced by the growing sophistication of publically available information on the Internet and elsewhere. Accordingly, this edition brings you several of these key examples where the power of new tools is evident and instructive.

To measure wetlands, and related landscapes, for this now expanded breadth of characteristics, the difficulty lies in developing novel and ingenious methods for collecting data as well as interpreting the data. Most research efforts conducted in a laboratory or in experimental plots seek to control all variables but one, and study different levels of the selected variable of interest or the large forcing function or stressors (Figure 2.12). Alternatively, landscape scale analyses focus on control of a given variable or a few variables, which can otherwise be difficult and potentially expensive.

FIGURE 2.12
A low-altitude look at a pothole lake in the glaciated Midwest. The surrounding land cover can be distinguished as can plant residue and their shadow along the northwestern shore of the lake.

Therefore landscape scale studies have a key advantage in that they often utilize survey sampling techniques (Congalton and Green 1998, 2009) where large sample sizes address a number of potentially associated variables. This can be a successful approach where the typical empirical approach may not be feasible. This edition will demonstrate the advantages to using multiple factors to develop correlative analyses, allowing for the postulation of relationships between ecological and other physical parameters in the landscape.

Although the new wetland landscape characterization paradigm involves the use of advanced technology and methods, tremendous value also resides in the more traditional approaches. In this edition, we meld the two schools of thought, respecting the traditions of prior techniques and complimenting those techniques with the newer methods. To obtain the best quality and variety of data with a landscape scale often means using traditional survey sampling and other appropriate measurements and data organization technologies.

Collection of data from a distance is *remote sensing*, which has demonstrated its value for a number of applications. The variables of interest may be directly measured over large areas using uniform data collection methodologies, and may involve airborne or orbiting platforms, or other types of sensors located in the field and linked by telemetry such as a flux tower. Indirect variable or surrogate variables may also be measured to take advantage of the capabilities of remote sensing. Examples of direct measurements would be the inventory of general wetland types. Indirect measurements would be the location of different general wetland types, their presence and absence, their interspersion, their juxtaposition and perhaps fragmentation (Robinson et al. 1992) as compared to the known habitat requirements of biota too small to resolve given a remote sensor.

Because copious amounts of data are typically utilized in landscape scale evaluations, such data collections must be organized with tremendous human interpretation or augmented by computer analyses. The advent of spatial databases and GIS techniques make possible the storage of data in such quantities. The spatially based storage of data as feature locations (i.e., point coordinates) or boundaries (i.e., vectors), or other feature characteristics in raster formats (i.e., grid cells) makes implicit the spatial location of features and their associated characteristics. GIS allows retrievable storage and maintains spatial fidelity and quality assurance, and can accommodate as much data as are necessary to solve any wetland-related problems.

Current and future efforts are devoted to geospatial modeling of processes using GIS and remote sensor imagery systems, and thus a good understanding of the capabilities of these technologies is a critical component for the wetland landscape professional, which this edition brings to the reader in the form of projects and data examples.

Current Global Perspectives and Systems Approaches

The diversity of technology and infrastructure to monitor environmental systems (Figure 2.13), from global to local scales, is growing rapidly as public and private organizations increase their investment in them. Resource planning, decision making, and management requires environmental data that specifically answers the operational details of new real-world problems, which are now more routinely discovered and understood by the public, partly due to developments in visual media for conveying these data (i.e., imagery from maps, models, and processed airborne or satellite remote sensing data). Therefore good environmental planning and management requires excellent data to ensure compliance and to monitor for project or program successes. Remote sensing products (including aerial photography, and airborne and satellite sensor imagery) are increasingly useful for monitoring and reporting these requirements, established by national policies and international treaties, conventions, and agreements (Backaus and Beule 2005),

FIGURE 2.13
Wetlands can be found in the most unusual places where water and land meet. Here is a discharge channel into a river with an abundance of new sediment deposits as yet uncolonized by vegetation. The juxtaposition of light toned sands or sediment, and darker toned examples show the quenching or darkening influences of water and sediments.

and are indeed now common tools shared worldwide. With the advent of Google Maps, Bing Maps, Google Earth, and other easily accessible and user-friendly geospatial data viewers, the public at large now expects these data to be of high caliber and available to them for day-to-day decisions.

Remote sensing is used at regional scales to improve land management policies and associated decision making processes, especially in areas that are highly agricultural and that are urbanized (Miller and Small 2003; Bryan et al. 2009). For example, data from AVHRR (Advanced Very High Resolution Radiometer), MODIS (Moderate Resolution Imaging Spectroradiometer), and MISR (Multiangle Imaging Spectroradiometer) satellite sensors have been used to monitor biomass burning in Brazil (Koren et al. 2007), which has a global impact on all of society. A combination of historical aerial photography and satellite imagery can also be used to establish historic baseline maps for land cover, and to project the impacts of different land management policies through the application of different possible alternative futures scenarios designed by local decision makers and community members (Baker et al. 2004; de Leeuw et al. 2010) worldwide. Satellite sensors can also be used to monitor and assess important and incremental ambient or diffuse globally significant environmental conditions, such those related to air pollution (Engel-Cox and Hoff 2005) and urbanization, and the risks associated with those conditions (i.e., human health, or the potential for impacts to people and built structures from flooding or fires) (Miller and Small 2003). Satellite platforms that are proving to be especially useful for the management of urban areas with regard to environmental risks around the globe include the relatively fine resolution IKONOS and Quickbird satellites, as well as the National Aeronautics and Space Administration's (NASA) Shuttle Radar Topographic Mapping (SRTM) mission (Miller and Small 2003).

For globally relevant and impactful issues, environmental dynamics may only be understood by appropriately scaled remote sensing data (Running et al. 2004; Reid et al. 2010). For example, satellite imagery from the Total Ozone Mapping Spectrometer (TOMS) sensor measures and reports the depletion of the stratospheric ozone layer over earth's poles, confirming previously hypothesized levels of ozone, and its declination (Molina and Rowland 1974; Engel-Cox and Hoff 2005; De Leeuw et al. 2010). At present, remote sensing technology is used to track global ozone layer holes on a daily basis by a partnership between the U.S. National Oceanic and Atmospheric Administration (NOAA) and NASA (NASA 2011).

The initiation of large-scale programs such as the Global Monitoring for Environment and Security (GMES) for the European Union (Backhaus and Beule 2005) and the National Ecological Observatory Network (NEON) in the United States also highlight the growing importance of remote sensing, spatial databases, and global scale datasets. NEON makes heavy use of remote sensing information, using an Airborne Observation Platform (supporting LIDAR, digital photography, and imaging spectrometer sensors) to observe regional trends in land use change, invasive species, and ecosystem functioning (Kampe et al. 2010). This is all in tandem with extensive in situ monitoring and data collection, all to leverage modeling for a better understanding of processes (Figures 2.14 and 2.15).

The Global Earth Observation System of Systems (GEOSS) aims to combine both remote sensing data and ground-based data into a meta-database that could be used to identify global environmental risks and evaluate management and policy successes at the global scale (Herold et al. 2008; Stone 2010). GEOSS provides information and tools for nine areas: disasters, health, energy, climate, agriculture, ecosystems, biodiversity, water and weather (Christian 2005; Lautenbacher 2006; Group on Earth Observations 2011). The development of GEOSS is driven by the Group on Earth Observations (GEO), an international consortium with members from 85 countries and numerous international organizations (Stone 2010). GEOSS is one example of the growing movement to release large data sets to the public, such as NASA's release of all historic Landsat imagery and the European Space Agency's intent to make all future satellite imagery free and open access (Stone 2010). Naturally the goal is to create a system of systems and supply data and tools to a larger audience of the public and decision makers.

Throughout the world, remote sensing data have been used extensively to assess both the loss of wetland area and, to a lesser degree, to assess the loss of wetland functions and services. These efforts have led to a more detailed understanding of wetland conditions, such as vegetational community

FIGURE 2.14
A series of images can provide examples of discharge, such as waters known as acid mine drainage areas in Athens County, Ohio. A pair of photographs (also see Figure 2.15) illustrates this location and this example is natural color or visible spectrum image, rendered in black and white for publication.

FIGURE 2.15
The corresponding image in Figure 2.14 completing the pair is a color infrared image rendered in black and white. Color infrared photography has two emulsions that are sensitive to the green and red reflectance of the earth, and a near-infrared sensitive emulsion. Typically, healthy, vigorously growing vegetation is very reflective in the near-infrared but dark toned in the visible spectra and hence appears dark in tone. Dead vegetation is very reflective in the visible and moderately reflective in the near-infrared and hence is shown here as very light toned areas near the highway. The dead vegetation is no doubt a victim of acid mine drainage.

characteristics, hydrologic regimes, and soil conditions. These biophysical details can identify degradation that affects the total functional area of a wetland, as well as the ecosystem services it may provide. However, these efforts have not been used to create a comprehensive global wetland assessment.

The work on wetland assessments in particular countries, such as the United States may serve as a template (or pilot) for global assessments of the future. Four decades of remote sensing data from several series of global orbiting satellites offer ample opportunities to establish baselines for global assessments of wetlands (de Leeuw et al. 2010). Habitat conservation and management applications, for example, have great potential for the widespread applicability of remote sensing data and analysis technologies as well as applications for predicting risks to water resources.

Ecological Goods and Services

Currently, national and regional surveys of wetlands assess the distribution and ecological conditions of wetlands and, much less commonly, the ecological functions of wetlands. To assess wetland ecosystem services for these relatively broad geographic extents it is necessary to link ecological functions with the production of ecological goods and services. This is a laudable goal and is a major challenge currently facing environmental and resource management professionals.

In the United States, indicators and the associated metrics are being used in state and regional assessments of wetland conditions as well as for the National Wetland Condition Assessment (NWCA) (http://water.epa.gov/type/wetlands/assessment/survey/index.cfm; accessed July 21, 2012), which utilizes a combination of field-based data and remotely sensed data. Current plans in the United States are to integrate the NWCA data from a variety of scales, from local to regional to national, and to quantify the provisioning of wetland ecosystem functions and services. However, the techniques for broad-scale mapping of ecological goods and services are in their infancy, and the connection between ecological functions of wetlands and the ecosystem services they provide is one of the potential impediments to the success of these efforts.

Landscape characterization, landscape ecology, and mapping (i.e., landscape metrics and landscape indicators) for determining relationships between abundance, distribution, and function of wetlands in the landscape, and associated delivery of ecosystem services are vital to advancing the science and understanding. It is important to note that the landscape approach does not preclude collection or use of field-based data, which may be a critical part of mapping wetland ecosystem services. In fact it is an encouraged activity and a necessary part of exercises. Much like calibration and validation of model algorithms as they mimic nature, so too is it necessary and desirable to conduct validation of wetland studies.

Quantitative linkages between field data and broad-scale landscape data are important for calibrating remote sensing and some GIS data processing steps, and are critical components of model scaling and validation. Integration of landscape characterization and modeling with ground-level wetland functional assessments is a key part of linking processes. This is also critical to developing ecosystem function and service models and maps that are relevant to decision makers at multiple scales.

Functional classes of wetlands can be related to the elements of ecological functions for wetlands (Heilman et al. 2002; Forman et al. 2003; Coffin 2007). Further, the changes in the relative abundance of wetland functional classes in a landscape can influence wetland ecological performance functions and related services (Lugo and Gucinski 2000). The use of the *landscape profile* concept (Gucinski and Furmiss 2001) may be one of several, initial steps used in the United States to scale-up the mapping of ecological functions of wetlands, as they affect the delivery of global wetland ecosystem services.

Some necessary actions, which can be perceived as important research needs for a better understanding of relationships between abundance, distribution, and functions of wetlands in the landscape, and the delivery of global ecosystem services have been developed. They can include:

1. Determining how results from wetland condition assessments can be used to estimate the delivery of ecosystem services by wetlands
2. Determining how wetland functions are associated with landscape characteristics, and how the distribution of functional classes of wetlands affects the delivery of ecosystem services
3. Developing landscape approaches (i.e., landscape profiles, functional surfaces) for determining hydrologic and ecological functions of wetlands and associated delivery of ecosystem services
4. Developing landscape models predicting delivery of specific ecosystem services and bundles of services based on wetland landscape profiles, empirical stressor-response models, and published literature

Although the characterization of wetland ecological goods and services is in its infancy, some pioneering work has begun. In a globally reaching and seminal effort to synthesize the extent and scope of all ecosystem services, the Millennium Ecosystem Assessment (MEA 2005) has provided tremendous leadership in summarizing ecosystem services (used interchangeably with *ecological goods and services*). Included in the multinational and interdisciplinary MEA is a detailed synthesis of water and wetlands, which has also brought wetland ecosystem services to the forefront of global environmental resource management. This approach is resulting in the clear articulation of the many fundamental ecosystem services provided by wetlands.

Wetland Ecosystem Supporting Services

Carbon Cycling

Wetlands are significant carbon reservoirs and contribute to ameliorating global climate processes through sequestration and release of fixed carbon. Carbon is contained in the standing crops of vegetation and in litter, organic soil, and sediments. The magnitude of storage depends upon wetland type and size, vegetation, the depth of wetland soils, groundwater levels, nutrient levels, pH, and other biogeochemical factors, necessitating analyses of such factors. These carbon reservoirs may supply large amounts of carbon to the atmosphere if water levels are lowered or land management practices result in oxidation of soils (Figure 2.16). Because wetlands serve as significant carbon sinks, the destruction of wetlands will release carbon dioxide. Although much is still not understood about the role of wetlands in the global carbon cycle, it is generally agreed that the gain and loss of wetlands is a major input to global carbon budgets, thus affecting the amounts of carbon dioxide and other greenhouse gases in the atmosphere, with potential for contributing to climate change.

Wildlife Habitat

Wetlands are a reservoir for biodiversity, which has many links to human well-being. Estuarine and marine fish and shellfish (see following "Wetland

FIGURE 2.16
It should be recognized that wetland plants seem to grow wherever there is presence of hydric soils and water, which may be diagnostic in the field and from imagery. Seen here in West Virginia, these cattails are growing in fine soils in a reclaimed area.

Ecosystem Provisioning Services" section), various birds, and certain mammals require coastal wetlands to survive. Many of the U.S. breeding bird populations (including ducks, geese, woodpeckers, hawks, wading birds, and many songbirds) feed, nest, reproduce, and raise their young in wetlands. Migratory waterfowl use coastal and inland wetlands as resting, feeding, breeding, or nesting grounds for at least part of the year.

Wetland Ecosystem Regulating Services

Wetlands protect human well-being by mitigating floods and buffering the effects of coastal storms. Wetlands reduce flooding by absorbing rainwater and by slowing the downstream flow of floodwater. Wetlands also function as natural attenuators of water flow, slowly releasing surface water and groundwater. Trees, root mats, and other wetland vegetation also slow the speed of flood waters and distribute them more slowly over a floodplain, lowering flood heights and affecting erosion and accretion in the surrounding landscape. Wetlands decrease the area of open water (fetch) for wind to form waves, which increases drag on water motion, thereby decreasing the amplitude of waves or storm surges. The value of wetlands to reduce impacts of floods and storms has often been retrospective, based on the estimated costs of damage or loss after a flood or storm has occurred.

Wetland Ecosystem Provisioning Services

Many wetlands contribute to recharging groundwater aquifers that are an important source of drinking water and irrigation. Plants, microbes, and soils in wetlands affect water quality by removing excess nutrients, sediments, and toxic chemicals, therefore improving water quality for humans and aquatic biota. Wetlands intercept surface-water runoff from higher dry land before the runoff reaches open water, reducing eutrophication in downstream waters and preventing contaminants from reaching groundwater and other sources of drinking water.

Many of the nation's fishing and shellfishing industries harvest wetland-dependent species. Most commercial and game fish breed and raise their young in coastal marshes and estuaries. Menhaden, flounder, sea trout, spot, croaker, and striped bass are among the more familiar fish that depend on coastal wetlands. Shrimp, oysters, clams, and blue and Dungeness crabs likewise need these wetlands for food, shelter, and breeding grounds.

Wetland Ecosystem Cultural Services

Wetlands also have recreational, historical, educational, aesthetic, and other cultural values that are held by the public and vary by geographic area and the societal values of the particular population of humans in an area.

Sustainability

The pervasive use of the term *sustainability* has, similar to ecosystem services, swept the globe, particularly in the past decade. The concepts embodied within sustainability are not new; however, their current incarnation unifies social and ecological perspectives of nature and make possible a new and positive perspective so that societies have the capability to meet the needs of the present generation while living within the carrying capacity of existing and supportive ecosystems, all the while providing future generations with opportunities to meet their own needs. This viewpoint and concept has had a tremendous effect on the need for quantifying ecosystems, such as wetlands (i.e., identification and characterization), their condition (i.e., ecological functions), and their impacts on society (i.e., ecosystem services) (Reid et al. 2010).

From the most general concept of sustainability comes a broad view of environmental and ecosystem management issues, which may offer a way to go beyond technological solutions to environmental problems by integrating social participation and policy dialogue with ecological inventorying, monitoring, and assessment activities. These concepts were codified by many of the world's representatives to the 1992 United Nations Conference on Environment and Development who provided us with an impression of the tremendous magnitude, promise, and challenges that sustainability can bring to the management of the world's natural resources.

The globally emerging emphases on sustainability bring with them a strong message of harmony with nature and with one another. These concepts also bring with them a number of other aspects of the appreciation of nature, depending upon the cultural or other values of a community including a variety of concepts. These can be associated with value systems that consider the immutable rights of animals and plants as well as the sacredness of physical sites throughout the natural world. No doubt that these messages that are part and parcel of sustainability (that is, the universally appealing rhetoric of being a caretaker of nature and living in harmony with nature) have potential for encouraging and facilitating the efficient use of resources but offer up a tremendous diversity of opinions due to the ever-increasing speed and intensity of cultural homogenization. Nevertheless, although a matter of degree depending upon particular cultural values, the differential recognition of the pure rights of the environment has universal appeal and, thus, global significance. Such approaches to environmental and ecosystem management would require preserving the local environmental values held by individuals, including all or a combination of several spiritual, cultural, and personal beliefs, while maintaining a close handle on the global (cumulative) impacts.

As alluded to previously, the current era, which has now been deemed by some as the *Homogecene* (i.e., resulting from the homogenization of people and societies), provides us all with numerous challenges for recognizing these different value systems when making environmental management decisions. By carefully melding a multitude of global value systems, but preserving and respecting diverse cultural and religious value systems that exist throughout the globe, we increase the likelihood of harmony with nature and each other.

In a world that is trending toward homogenization of its people, cultures, languages, and ecological systems and functions, this global path to sustainability, which includes cultural values, may be extremely difficult without intentional efforts. Agreements are necessary to engage in the intercultural or interfaith dialogues about the sustainable use or alternatives to use necessary to be successful resource managers, planners, and decision makers. There are numerous examples of global environmental issues that are truly impending crises. Hence, they must be dealt with thoughtfully and effectively if societies are going to live in true harmony with nature, a fundamental part of sustainability. A delicate balance of globalism and inclusion of critically necessary local and regional perspectives may be required to develop robust, adaptive, and respectful formulation of natural resource management plans. Naturally, wetland ecosystems are an important part of that planning.

Lesser developed countries face the challenge of reducing poverty and attaining sustainable development at the same time. They have a particular need (and incentive) to benefit from this unifying concept at a global scale. A key link in this global concept and perspective is the wise use of geospatial information, particularly for natural resource characterization and other assessments, including wetlands, due to their ubiquity and importance in delivering crucial ecosystem services.

The emerging critical and future threats to wetlands and related natural resources, and the sustainability of ecosystems are soil loss and degradation, water scarcity, and the loss of biological diversity (Running et al. 2004), regardless of the sociological contexts. The perceptions of these environmental problems vary tremendously, depending upon a number of socioecological factors. If geospatial information is to have an impact on the users in these areas, the information they produce needs to be compelling, accurate and easily accessible to the user (i.e., must have high impact and availability). Some argue for an approach that addresses this complexity as a *multilevel stakeholder approach to sustainable land management*, for finding feasible, acceptable, viable, and ecologically sound solutions at local scales. A number of international programs and bilateral cooperation projects have also taken this perspective and started using these *sustainable land management* (SLM) approaches, either explicitly as in the case of United Nations Capital Development Fund or at least implicitly in their other programs. The SLM approach uses management as an activity on the ground, with appropriate technologies in the respective land use systems. This SLM sustainability paradigm requires that a technology follow five major sustainability principles: (1) ecological protection, (2) social acceptance, (3) economic productivity, (4) economic viability, and (5) risk reduction. Accordingly, a technological approach to resource management that is sustainable would have to be developed using criteria for a particular and locally relevant land use and would likely not be applicable everywhere. This SLM method follows on the earlier discussion in that it encourages the full exploration and consideration of all dimensions, particularly the economic, but also the social, institutional, political, and ecological dimensions of the community in question. Interestingly, for resource managers, this new local-to-global viewpoint of sustainability creates an entirely new dimension, which has been generally referred to as the realm of the *stakeholder*.

A number of global environmental professionals have also suggested the efficacy of explicitly linking research on global environmental change with sustainable development (Reid et al. 2010), which would necessitate the increased use of remote sensing and geospatial analysis for monitoring ecosystem conditions, and also for measuring feedback loops between environmental conditions and societal values and activities. Remote sensing data could thus be incorporated into large and complex models of socioeconomic systems to help determine the social and economic impacts of wetlands and other environmental components or changes. The data could also be integrated with existing efforts, such as GEOSS, so that its products would best meet the needs of decision makers about sustainable development issues.

Watershed and Coastal Planning

Flooding is a major natural hazard that impacts different regions across the world every year. Between 2000 and 2008, among the various types of natural hazards, floods have affected the largest number of people worldwide, averaging 99 million people per year (WDR 2010). Within the United States from 1972 to 2006, the property losses due to a catastrophic flood event (a flood causing damage of $1 million or more) averaged approximately $80 million (Changnon 2008). On average, floods kill approximately 140 people each year (U.S. Geological Survey 2006).

In 2011, in the United States, the Mississippi River experienced unprecedented flooding that caused fatalities, evacuations, and large financial losses. Climate change is expected to continue enhancement of the risks of extreme storm events (Milly et al. 2002) and increases in the frequency of flash floods. Large-area floods in many regions are very likely to occur in the future (Alley et al. 2003; Parry et al. 2007).

In addition, the world is also undergoing the largest rate of urbanization in history. In 2008, for the first time in history, half the world's population resided in urban areas, and urban areas are expected to experience most of the future global population growth. From 1982 to 1997 the amount of land devoted to urban and built-up land uses in the United States increased by 34% (U.S. Department of Agriculture 2001). Urbanization can also increase the size and frequency of floods along river courses, and may also expose those communities to increasing flood hazards (U.S. Geological Survey 2003), increasing the future focus by planners and land managers on the role of urbanization in the prediction of flood levels and damage, primarily for disaster management and urban and regional planning (Milly et al. 2008).

A key example in the Midwestern United States, the Kansas River, has been prone to flooding over the last century. It has faced two significant flood events in 1951 and 1993. The damages from the 1951 flood were substantial. Nineteen people were killed and about 1,100 injured with total damage estimates as high as $2.5 billion (Juracek et al. 2001). During the height of the flood, on July 13, 1951, nearly 90% of the flow in the Missouri River at Kansas City came from the Kansas River, a tributary comprising only 12% of the Missouri's drainage basin (Juracek et al. 2001). The historic 1993 Midwestern floods affected nine states and resulted in $15 billion worth of flood damages.

An effective approach for assessing flood risks for people and their property within a watershed, such as along the Kansas River, is through the production of flood risk models, which show areas prone to flooding events of known return periods. To mitigate flood risk, reservoirs and levees have been historically utilized, and in recent years wetlands have been studied for their flood attenuation and water quality improvement potential (e.g., Mitsch et al. 2001). As described earlier, wetlands also have the capability of short-term surface water storage and can reduce downstream flood peaks, and benefit biological and ecological functions in the landscape, as well as wildlife habitat and water quality characteristics of the landscape (Hey and Philippi 1995; Lewis 1995; Wamsley et al. 2010).

In coastal areas around the world's oceans and lakes, millions of people live within kilometers of the coast, and make up a large percentage of the world's population. Therefore, extreme change is a critical concern for coastal planners and other decision makers, including developed areas that are intimately linked to surrounding tidal, estuarine, and other coastal wetland areas. The extent and the degree of the loss of coastal wetlands is an important and challenging question for geographers, ecologists, and policy makers alike, in order to adequately address the human and other ecological impacts from coastal land use and changing sea levels in coming decades.

From a tidal marsh management perspective, coastal wetlands are among the most susceptible of ecosystems to changes in sea level rise and associated changes in sea water surge intensity, and also from increased predicted surge heights resulting from hurricanes, typhoons, and tsunamis. Some of the impacts on wetlands from sea level rise and surge are a result of a wetland's particular sensitivity to short-term and long-term changes in inundation, salinity, and soil characteristics.

There are a number of landscape approaches increasingly being used to measure and monitor effects, for example, the predictive GIS tool, known as the Sea Level Affecting on Marine Marshes (SLAMM) model in coastal marshes, and other simpler models. In subsequent chapters we demonstrate the importance of considering in ones selection of modeling approaches to planning, the trade-off between information resolved by the approach (i.e., the resolution of the information), the availability and applicability of uniform and consistent data for input to a particular model, and cost of producing an answer that is suited to the decision-making needs. In other words, the scale of a management problem and resolution of information necessary to have an impact on associated decision making are the interactive drivers of project cost. This principle also applies to many other landscape level assessment projects and programs, which we will discuss in further detail in coming chapters.

Relatively coarse-resolution approaches using models such as those that simply utilize coastal land elevation data could provide adequate impact for decision making, especially at the continental or global scales. Relatively fine resolution, that is, more detailed approaches using models, such as those described later, may be more applicable to local and subregional analyses and coastal planning than the relatively coarser models. We provide details of these differences and their differential impact on decision making in later discussions.

Utilizing This Edition to Its Full Benefit

This edition is written for all environmental and resource professionals in the fields of engineering; ecology; resource management; climate sciences; and policy development at the local, state, regional, national, and global scales. It is likely a valued resource for those increasing their understanding of broad scale wetland science and geospatial technologies. Results and information address the linkages, and hence can inform issues related to national policies such as the Clean Water Act, No Net Loss Policy, and other policies with relevance to international commissions, conventions, protocols, or agreements that require distilling information from community- and place-based analyses, and applying relevant components of the finer scaled decisions. A worthy example, discussed later, is work ongoing in the transboundary ecosystem of the Laurentian Great Lakes, where Canada and the United States have successfully collaborated to monitor and assess the ecological functions and services of coastal and other wetlands, with tremendous strides in recent and novel research and restoration, proposed for funding in 2010 for $475 million. This edition of the book presents a transboundary solution to evaluating and monitoring Great Lakes wetlands along with the techniques that likely have general applicability to other large area projects and programs.

This edition also provides many applied examples with relevant detail using current techniques that address the needs of wetland landscape ecology practitioners. These are applied to the diverse wetland types, conditions, and issues encountered every day by practitioners. A number of resources from the literature are noted for reference, which provides the reader of this edition with a tremendous amount of integrated information that provides a time-saving and clear path toward solving specific challenges. Thus these methods and approaches provide new users as well as professionals with practical tools for success. This is particularly true of the present day complex situation of understanding, conceptualizing, designing, and implementing landscape scale wetland projects by providing the necessary detail and direction. With these approaches and methods one can understand and carry out one's exploration of the science and societal linkages involved.

Selected new areas included in this edition are

- Updates of practical geospatial methods
- Demonstration of specific examples, which are project driven
- Descriptions of any pitfalls of using ecological data at landscape scales, with solutions
- Discernment of alternative techniques for a variety of practitioners
- Inclusion of linkages between field and landscape ecological practices
- Online resources for practitioners

3

Traditional to Contemporary
Characterization Methods

Photointerpretation

Remote sensing technologies include a variety of tools ranging in complexity from advanced sensors to simple camera and film systems. Each imaging system creates two-dimensional image products that can be analyzed or interpreted for valuable information. The process of analyzing these images is called interpretation of images.

The generally accepted definition of image or photointerpretation incorporates several steps or elements, including the fact that interpretation is an art and science of obtaining or interpreting data from the characteristics of features recorded on photographs or images. Photointerpretation refers specifically to interpretation of photographic products, whereas image interpretation is more general as it covers the interpretation of all image products including photographic examples. Generally, these terms refer to interpreting images or photographs taken from aircraft or spacecraft. The methods used for interpretation are the same methods and skills that can be used to obtain information from any type of photograph or image. Interpretation applies several methods through skillful application to obtain details about features found in images. These elements of interpretation supply information on features that are basically independent assessments of each characteristic. These elements include characteristics such as color or gray tones, shape, size, texture, pattern, shadow, and associations (Figure 3.1).

The gray tonal range supplies detail about features or materials in a range of contrast between black and white. Typical materials exhibit a range of gray levels on visible black-and-white photographs. Rock, bare soil, concrete, and similar impervious ground materials often appear light toned. Vegetation is relatively dark toned due to the low relative reflectance of green plant material compared to bare soil or rock.

The variability of visible light can be measured through the use of color film. Colors show the tonal variation of each additive color primary, such as ranges of blue, green, and red. Color, like tone, can be used to evaluate

FIGURE 3.1
Forested wetlands can be identified from this low altitude aerial image. Note the central dark shape of the tree canopy. From this altitude, it is possible to identify the general tree community type(s), such as deciduous or evergreen, as well as the vertical structure of the forest. At a lower viewing altitude, identification of species is possible.

features as well as pinpoint the spectral variability of materials in the visible portion of the spectrum.

Shape refers to the exterior configuration of materials or features. Many features or objects have distinct regular or irregular shapes. Shape can be a unique clue as to the identity of the feature.

Size refers to the absolute or relative dimensions of the object or feature (Figure 3.2). Size can be very important as features are often indistinct on photographs and knowledge of size can be a big clue as to the identity of a given feature.

Texture is a variation in tone or color caused by a mixture of materials on a given site. Interpreters will identify a texture as a faint variation in tone or color. For example, in an old, abandoned farm field the texture may result from a tonal or color contribution from the presence of shrubs or small trees in the field. In the eastern United States, the trees may be sumac or aspen, and in the western or mountainous regions they may be small conifer trees or shrubs. These shrubs or trees are too small to be recognized by their individual crown or canopy vegetation. Yet they contribute to the overall variability of tone, as compared to the relatively uniform tone of a mature soybean or wheat crop in a field. The presence of tall plants, partial canopy closure, and shadow help create a textural difference that often allows the separation of corn crops from soybean crops.

Pattern, conversely, is the regular or irregular distribution of relatively large features on the earth's surface. An example is the regular pattern of

FIGURE 3.2
Low-altitude photographs can be very useful to delineators. A small digital camera, a rented aircraft, and a steady hand caught this image. Several contextual clues can be gleaned from the size of features in images, such as this one, supplemented by tone and texture information. Note the size of buildings and roadways, which can be used to gauge the width of mottled features in fields, or the width of fence rows, where trees and shrubs exist on the boundaries of agricultural fields.

field boundaries in a rural area. When this pattern is disturbed, for example, by a stream course, one can infer certain characteristics of the stream channel such that it is too big to be altered by human activities over time (Figure 3.3). Often these channels will be obscured by vegetation such that the water itself is hard to see. The pattern of the riparian vegetation still indicates the true characteristics of the feature and the presence of vegetation that shades the stream and offers habitat for biota.

The shadow of a feature can be an important clue and may help identify the feature regarding type. Shadows cast by a feature can provide additional important shape and size information. A common example is that of interpreting the name of a business or building from the shadows of individual letters that form a sign. The individual letters *blend* into the building, but the shadows are distinct to the interpreter or may be too narrow to view from a remote sensing platform, such as an aircraft or satellite. Conversely, shadows can obscure detail. It is difficult to view a feature hidden by a shadow, due to the lower relative illumination in the shadow area as compared to the overall illumination of the sunlit areas.

The *association* of a feature is the other characteristic or clue that is found together, in association, with the feature of interest. For example, the course of a stream is usually evident by shape, but often the tone of the water is hidden from view by trees. Stream courses are usually found in association with

FIGURE 3.3
The association of shadow and pattern of vegetation can again tell a story to the interpreter. Note the rough texture in the center of the image, with drainage ways. The right-to-left pattern is from row cropping, but the bottom-to-top pattern is that of seepage or drainage of fields leading to the woods. Note that this information provides clues that hydric soils have potential in this area, or there is a history of hydric soils.

other clues such as the meandering stream pattern, the branching drainage pattern, automobile and railroad bridges, pond or lakes, stream-side vegetation, lower relative elevation, and a downhill course as compared to the surrounding landscape.

Interpretations

Interpretations of images can supply a great level of detail on wetlands, water features, and associated land cover types. Use of the elements of image interpretation along with knowledge of the features of interest and experience at interpretations can provide valuable information.

When viewing images, interpretation supplies information on wetlands and aquatic ecosystems (Niedzwiedz and Ganske 1991). One can see details of the nearshore topography or bathymetry due to the penetration of visible

FIGURE 3.4
Capturing phenomena during an overflight is valuable and instructive. Here differences in
the tone of crops and underlying soils tell a story. Mid-image there are irregular patches of
dark-tone weeds among the light-toned crop. Likely these were wet spots at the time of plant-
ing, which resulted in seed kill and later growth of non-row-crop vegetation in the dark tones.
Could this wet or now dark-toned area have once had hydric soil conditions?

light into and back from the water. Often the shape of submerged, coastal
aquatic features can be identified due to the penetration of visible light into
the shallow water column. One should note the very light tone or color of
bottom sediments, roads, and beach areas (Figure 3.4).

In comparison, the vegetation of the submerged aquatic plants or submergent wetlands can be identified in shallow water areas due to the dark
relative tone or color as compared to the surrounding light-toned sand or
sediments. The contrast between light-toned or colored bottom sediments
and the relatively dark tone or color of submerged vegetation allows for the
identification of features of interest when making a resource determination
or characterization decision.

Interpretation clues can be articulated for a given ecosystem under study
and formalized into a list of characteristics or interpretation key to facilitate identification and inventory. This approach has worked well in a number of studies and is used to great effect to characterize submerged aquatic
vegetation (SAV) that is the focus of habitat studies for a number of species (Raabe and Stumpf 1995), resulting in, for example, a shared benefit to
the Chesapeake Bay restoration efforts in the United States (Ackleson and
Klemas 1987).

The use of differences in spectral reflectance as described earlier can facilitate a number of applications. Analysis of reflectance characteristics shows
that dark-toned features found beneath or above the water are emergent and

submergent vegetation. They appear dark because plants reflect little light in the blue part of the visible spectrum and in the red part of the spectrum. On a color or black-and-white image these plants will be dark tones or dark color, as the green reflectance of light by plants is relatively low compared to that of soil.

Interpretations are facilitated by the evaluations of differential spectral reflectance of a variety of portions of the electromagnetic spectrum. The use of a number of images and other different film types can be a great help in interpretation (Lyon 1987; Williams and Lyon 1991; Lyon and Greene 1992; Lyon 1993; Lyon and McCarthy 1995). This is particularly true of hydrological characteristics of areas that change over short or long periods of time. Multiple dates of photos or images, and multiple types of films or sensor band data can help identify the general hydrological characteristics of wetland areas by capturing wet, dry, and normal moisture periods.

Following rainstorms and other hydrological-related events, the water may be opaque due to the runoff of and resuspension of sediments. These suspended sediments or nonpoint pollutants may be difficult to image through and can disrupt evaluations of submergent wetlands (Lyon et al. 1988). Conversely, the suspended sediments may help to locate the movement of water and waterborne pollutants. Multiple images can help lend insight into a variety of conditions by capturing the variability over time.

It is always useful to obtain many multiple dates of aerial photos or remote sensor data of a given study area in support of analyses. This is because each photo supplies unique information and repetitive coverage adds to the value of multiple samples and statistical analysis capabilities to the project. Photos may also be inexpensive in comparison to the costs of obtaining the similar quality and quantity of field data using traditional methods.

Photos and images are also valuable in location of the position of features and materials. Knowledge of size and position can be supplied by the measurement capabilities of images (Figure 3.5). Analysis of images can provide horizontal and vertical information by using a combination of photogrammetry or surveying technologies. Photogrammetric measurements can be made on the images and tied to some absolute reference to characterize the location of earth or terrestrial features in an absolute sense. Surveying also allows one to later relocate these features using the original measurements or map products.

Historical Aerial Photographs, Sources, and Their Utilization

Another valuable source of data in planning and monitoring is historical aerial photographs. Aerial photographic coverage of the United States is available from archives. Generally, photographs date to the late 1930s. Multiple

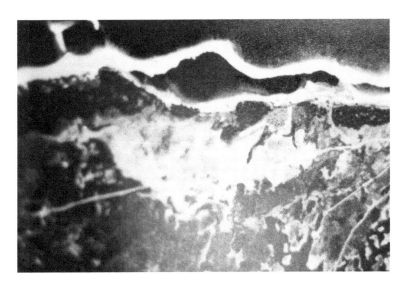

FIGURE 3.5
This relatively low-resolution medium-altitude aerial photograph demonstrates the value of using a combination of images to see the extent of resources and adjacent resources. Note the open water and its dark tone at the very top and the very light-toned beaches. The mix of gray tones and patterns results from mixtures of water, sand or soils, and vegetation. Analysis of these tones or colors can help one interpret wetland resources. Use of medium- and low-altitude images, such as the following examples, can lead to a convergence of evidence. Location: in the Wilderness State Park and Waugoshance Point on the northern lower peninsula Michigan near the Straits of Mackinac, which connects Lakes Michigan and Huron.

sets of aerial photos are in archives, and one may be able to develop a time series of photos extending from the 1930s to the present (Figures 3.6 to 3.8).

The interpretation of historical aerial photos provides data on a number of conditions. From individual dates of coverage, interpretation will yield data on the land cover types, presence or absence of houses and buildings, stream drainage pattern, and general soils and geomorphology characteristics. Multiple dates of coverage allow the user to capture the different hydrologic conditions of wetlands, lakes, and coastal areas that have occurred over time (Butera 1983; Lyon et al. 1986; U.S. Army Corps of Engineers 1987; Carter 1990; U.S. Environmental Protection Agency 1991; Williams and Lyon 1991; Lyon and Greene 1992; Lyon 1993).

Sources of historical aerial photographs at the federal level include the National Archives and Records Administration in Washington, DC, for pre–World War II photographs; the U.S. Geological Survey EROS Data Center in Sioux Falls, South Dakota, for U.S. Department of Interior agency photographs; and the Aerial Photography Field Office in Salt Lake City, Utah, for U.S. Department of Agriculture (USDA) agency photographs. The addresses for these sources are publically available at each agency's Internet site and

FIGURE 3.6
A low-altitude image of the beach barrier and back barrier marsh, and upland forests near the Straits of Mackinac. The sandy barrier is light toned with specks of dark tone that are beach plants. To the left is a lagoon of dark-toned water with sandy soil lagoon bottom materials that reflect, upwelling the light and leading to a mixed tone of shallow water over sand. The surrounding shoreline has very dark tones, which are wetland vegetation obscuring the bottom with its canopy of submergent and emergent wetland plants.

in Lyon (1993), Lyon and McCarthy (1995), Ward and Elliot (1995), Ward and Trimble (2004), and other sources.

Aerial photographs or remote sensor images are usually available for every other year or every third year in the U.S. Geological Survey (USGS) and USDA archives. It is possible to gather approximately ten or more dates of aerial coverage since the 1930s from the aforementioned sources (Lyon 1981; Lyon and Drobney 1984).

Another good source of aerial photographs is the U.S. Department of Agriculture Farm Service Agency (FSA), formerly the Agricultural Stabilization and Conservation Service. Since approximately 1981 the FSA has collected small format (35 mm) color transparency aerial photographs of farmed areas subject to crop support programs (Lyon et al. 1986). The photos were taken on color slide film and acquired at relatively low altitude creating large-scale photos for analysis. These photographs have been archived within each state since approximately 1983, and recent coverage is archived in county FSA offices where they can be viewed and ordered. As programmatic functions and emphases change over time, it is always recommended to check with each office prior to visiting to obtain current availability.

Significantly, these photos have been acquired during the growing season of crops and other vegetation. Aerial mapping photographs are usually

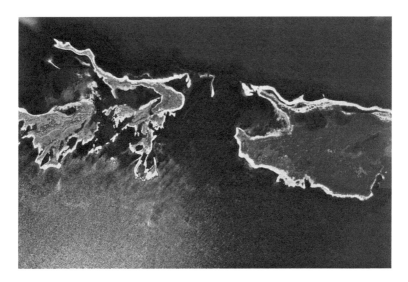

FIGURE 3.7
Back up to a medium-altitude image, one can see the mixture of light, gray, and dark tones that corresponds to the mixture of water, soil, and plant materials that make up lacustrine coastal wetland. Water extinguished light over depth and is dark toned or optically deep; a presence of shallow bottom of highly reflective material such as beach sand creates the tone that relates to the abundance of plant material.

FIGURE 3.8
A very low-altitude image from a small aircraft yield a close-up view of resources familiar to the bird's-eye view of the reader. Note the dark tone of tree vegetation and the characteristic pattern of evergreen trees and their shadow. Note the very light tone of the shallow bottom of the lake and its very light tone marl soils, quenched by a thin layer of water.

regularly acquired during the dormant or *leaf-off* period of vegetation. Leaf-off aerial photographs are valuable so that the stereoplotter operator can follow terrain contours and make topographic maps without the land surface being obscured by leafy vegetation (U.S. Army Corps of Engineers 1993). Because the FSA and other agency image coverages, photographs are acquired in the growing season, the wetland and aquatic vegetation is in active growth, as compared to the spring or fall period of leaf-off conditions that are usually specified for the acquisition of mapping photographs.

It is still possible to identify and inventory wetland areas using leaf-off photographs. This is because the extent of the wetland area is defined by the vegetation residue present from previous growth and from the difference in wetland appearance as compared to adjacent terrestrial or aquatic systems (Lyon 1993; Lyon and Lyon 2011).

Creation of Image-Based Land Cover Products

The demand for land cover products is driven by a deep need for contextual landscape information. In the past, land cover classifications and map products were customized in accordance with the project goals or scope using a team of photo interpreters lead by an expert. The team used the mentioned data sources to map land cover polygons according to a scope, and conduct the interpretation and classification process to produce the desirable land cover types.

To maximize effective use, polygons can be interpreted and marked on transparent overlays of high-altitude photographs. Polygons should meet the minimum mapping unit requirements described in a scope, and be drawn and identified using the land cover classification types.

A member of the interpretation team should work on the individual quadrangle areas and photos. The overlays can be made of polygons and of their land cover type. Work should be tracked by photointerpretation, containing information in numerical and in written form. Particular attention should be taken to assure correct land cover classifications. This includes training and monitoring the work of the team by the team's photointerpretation expert (Congalton and Green 1998, 2009).

Specific attention should also be paid to identifying wetlands using guidance from a scope and using the techniques described by Lyon (1993) in *Practical Handbook for Wetland Identification and Delineation*, which provides a number of valuable methods for inventory and classification of wetlands. Included are descriptions of methods to conduct a large-area or regional evaluation of wetland land cover types.

A quality assurance and quality control plan should be in place to track the photointerpretations by photo and quadrangle using the worksheets. Each overlay of a given image and its worksheet should be evaluated for the positional quality and classification accuracy by the expert. In this manner overall quality is evaluated by the expert and any problems are identified quickly. Solutions to problems should be recorded and distributed to the image interpreter for discussion and action (Congalton and Green 2009). The resulting GIS overlays can then be transferred photogrammetrically to the mapping base.

Applied Photogrammetry and Image-Based Remote Sensing

A variety of problems can be identified and monitored using aerial photographs, and airborne or satellite-based digital imagery. The fundamentals include the use of one or more images for interpretation or comparison purposes.

The differences in light reflected from one kind of land cover or land use as compared to another may be used as an indicator of what it is (i.e., interpreting the materials on the ground, from above) and potentially what it is if change has occurred (Lyon et al. 1998). The differences in gray tone in black-and-white images (or color, in color or color infrared images) can therefore be interpreted using static images or with multiple images of the same place in different weeks, seasons, years, or decades.

The difference in tone or color can be demonstrated or quantified from digital data. When a forest exists, the tone can appear dark on black-and-white images and on color images the addition of green to this dark area adds context for the interpreter. When houses are constructed among or nearby the forested wetland or marsh, the tone during construction is light on black-and-white images and the color is brown-white on color images. On color infrared (CIR) films or images bare soil areas appear blue-green in color and light toned or bright. These differences are very obvious under most conditions. The differences can be interpreted over time as construction of houses or buildings and as an exposure to wetland receptors.

Other characteristics and conditions of property can be of interest. Indicators of these conditions can be used to identify and locate areas for fieldwork and contribute to management of wetlands.

Periodic monitoring with aerial photographic, aerial videographic, and aerial or satellite remote sensor technologies can supply necessary periodic monitoring data. Remote monitoring is a benefit too, because it can be accomplished at low relative costs. Various imaging technologies have certain capabilities and costs. Orthophotography, for example, provides an excellent capability for producing large-scale map sheets. However, it can be relatively expensive for certain applications compared to other traditional data acquisition methods such as field sampling and field mapping.

There is tremendous value to the photointerpretive potential of historical and current aerial imagery and other data, and their utilization is where a demonstrable difference can be made to impact wetland issues and goals. Numerous examples of their use have been demonstrated. A leading example of the use of wetland monitoring for state policy and decision making is in the Midwestern state of Ohio. Current wetland water quality standards, criteria for wetland mitigation and restoration, and innovative policy steps make the work in Ohio a leader in this realm.

Case Study: Traditional Characterization of Geographically Isolated Wetlands

General patterns of ecosystem response to landscape change have been postulated (e.g., Harris 1984; Odum 1985; Opdam et al. 1993; Forman 1995), but such relationships have usually been studied in terms of an ecosystem's response to environmental extremes rather than as a response to a gradual transition in land cover or patchiness. In this case study, we used traditional photointerpretive techniques to examine the relationships between ecosystems and landscape change by describing the full range of ecological variability among thirty-one depressional wetlands (Brinson 1993) in central Ohio using photogrammetric interpretation. Similar approaches have been described and used in wetland plant communities potentially influenced by landscape change (Leibowitz et al. 1992; Peterson et al. 1996; Fennessy, Geho, et al. 1998; Fennessy, Gray, et al. 1998).

Correlations between landscape stressor and aquatic plant community composition have also been explored with a *gradient analysis* technique in the lakes of Nova Scotia and the Pyrenees Mountains by measuring stressor gradients among sites, thus relating landscape conditions to the structure of nearby biological communities (Catling et al. 1986; Gacia et al. 1994). However, gradient analyses require a priori pairing of stressor-variables with response-variables to be ecologically meaningful. The a priori pairing approach may obscure other relationships between the biological components of the landscape and the embedded ecosystem, particularly when several parameters are simultaneously compared (Yoder 1991; Karr and Chu 1997). Thus, as an alternative some biologists have measured specific plant community composition parameters, for example, percent native species (Wilhelm and Ladd 1988; Adamus and Brandt 1990), to ensure that meaningful ecological parameters are detected during gradient analyses.

Prior research suggests three predominant factors that are responsible for many ecosystem disturbances, all of which can be determined from remote sensing: (1) land cover type, (2) distance between ecosystems patches, and (3) ecosystem size (Godwin 1923; MacArthur and Wilson 1967; Simberloff and Wilson 1970; Diamond 1974; Nip-van der Voort et al. 1979; Moller and

Rordam 1985; Jones et al. 2001). In this case study we used aerial photo-graphs to develop plant-based indicators of wetland disturbance, focusing on depressional wetlands (Brinson 1993), which tend to be geographically isolated within the larger landscape (Leibowitz and Nadeau 2003). Wetlands ranged from highly impacted due to disturbance (e.g., wetland remnants on farmland) to less impacted (e.g., wetlands within nature preserves) in this case study. Analyses involved comparing a gradient of wetland plant characteristics among the thirty-one wetlands to the gradient of landscape characteristics of local land cover, interwetland distance (i.e., wetland frag-mentation) and wetland size (i.e., area).

Study Sites

Thirty-one depressional wetlands in central Ohio were selected as study sites (Figure 3.9) based on National Wetland Inventory (NWI) maps, aerial photographs, topographical maps (scale = 1:24,000) and county-level soil survey maps. Mean wetland size was 2.0 (SD = 1.7) hectares, ranging from 0.1 hectares to 8.0 ha, comprising a total wetland area of 63.5 ha. Sites were selected such that they generally spanned the gradient of current landscape conditions (Table 3.1) in central Ohio and represented emergent, scrub-shrub, and forested wetlands (after Cowardin et al. 1979). Five of the wetland study

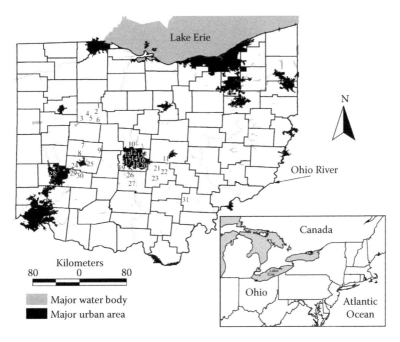

FIGURE 3.9
Location overview of thirty-one depressional wetlands studied in eleven Ohio counties, identi-fied by site number.

TABLE 3.1

Areal Percent of Land Cover Surrounding 31 Depressional Wetlands in Central Ohio

Wetland Site	Wetland Class	Urban	Agriculture	Grassland	Forest	Open Water	Bare Soil
			Areal Percent				
1	Forested	6.9%	62.7%	0.9%	28.4%	0.0%	0.0%
2	Scrub-Shrub	2.7%	74.3%	0.6%	22.0%	0.2%	0.1%
3	Emergent	2.3%	92.5%	0.0%	5.1%	0.0%	0.0%
4	Scrub-Shrub	0.3%	47.8%	0.0%	50.6%	0.0%	0.0%
5	Emergent	0.7%	59.4%	0.0%	39.7%	0.0%	0.0%
6	Scrub-Shrub	4.6%	42.8%	3.5%	48.8%	0.0%	0.0%
7	Forested	3.6%	53.6%	4.7%	37.9%	0.1%	0.0%
8	Emergent	11.5%	71.0%	0.7%	14.0%	0.8%	0.0%
9	Scrub-Shrub	5.3%	82.0%	0.0%	11.5%	0.9%	0.0%
10	Emergent	24.6%	0.0%	13.2%	57.9%	1.8%	0.0%
11	Forested	21.3%	55.8%	0.8%	18.8%	0.8%	0.0%
12	Forested	90.9%	2.7%	1.0%	3.6%	0.1%	0.0%
13	Emergent	29.8%	65.5%	0.3%	4.2%	0.0%	0.0%
14	Emergent	59.1%	0.0%	13.7%	26.2%	0.3%	0.0%
15	Forested	57.2%	0.0%	4.8%	37.9%	0.0%	0.0%
16	Emergent	53.0%	3.9%	4.5%	38.6%	0.0%	0.0%
17	Emergent	78.1%	0.0%	0.0%	17.3%	4.2%	0.0%
18	Emergent	78.1%	0.0%	0.0%	17.3%	4.2%	0.0%
19	Emergent	20.8%	66.7%	0.0%	12.5%	0.0%	0.0%
20	Emergent	17.1%	70.3%	0.7%	11.4%	0.0%	0.0%
21	Scrub-Shrub	3.8%	65.2%	0.0%	30.4%	0.6%	0.0%
22	Emergent	4.9%	64.2%	0.0%	30.2%	0.6%	0.0%
23	Forested	2.2%	79.6%	0.0%	18.0%	0.0%	0.0%
24	Scrub-Shrub	6.5%	82.6%	0.0%	10.2%	0.0%	0.0%
25	Scrub-Shrub	1.6%	85.9%	0.0%	12.0%	0.0%	0.0%
26	Emergent	6.9%	61.2%	10.1%	15.1%	6.0%	0.0%
27	Emergent	3.3%	83.1%	0.0%	10.8%	0.0%	0.0%
28	Forested	76.1%	0.0%	0.0%	19.0%	3.9%	0.0%
29	Emergent	17.8%	44.5%	3.5%	26.5%	1.2%	6.0%
30	Emergent	27.3%	35.4%	6.1%	23.2%	1.3%	5.8%
31	Scrub-Shrub	0.7%	0.0%	9.6%	88.8%	0.0%	0.0%

Note: Wetland class per Cowardin et al. (1979).

sites underwent restoration or construction prior to this study: (1) Site 10 was constructed in the early 1990s; (2) Site 14 was constructed in the early 1990s; (3) Site 17 was restored from row crops to wetland in the early 1990s; (4) Site 18 was restored from row crops to wetland in the early 1990s; and (5) Site 29 was restored from row crops to wetland in the mid-1980s. Initial analyses indicated that there were no significant differences between the five restored or constructed emergent wetlands and the remaining eleven emergent wetlands with regard to mean aboveground biomass production, taxa richness (i.e., species, genus, or family), or the presence of plant species typically observed in disturbed areas. Thus, the five restored or constructed wetlands were included in the gradient analyses. Observations during the three study years indicated that the thirty-one wetlands underwent no direct disturbances that would impact the study results. Ambient environmental conditions in Central Ohio include mean annual precipitation of 95 cm, with mean temperatures ranging from a high of 72°F (22°C) in July (70% humidity) to a low of 37·F (−3°C) in January.

Vegetation Assessment

The vegetation of the thirty-one wetland study sites was sampled in either July or August, with three sites sampled in 1996, twenty-one sites sampled in 1997, and seven sites sampled in 1998. Prior to vegetation sampling, wetland boundaries were determined using hydrologic indicators (Lyon and Lyon 2012) and the relative presence of wetland flora (Reed 1988) at wetland boundaries. A nested quadrat sampling method (Mueller-Dombois and Ellenberg 1974) was used to sample herbaceous plants, shrubs, and tree species at each of the wetland sites.

Approximately thirty 0.45 m² circular quadrats were sampled along two perpendicular transects that passed through the approximate center of each wetland. Exceptionally small wetlands were sampled with as few as twenty-two quadrats (Sites 7, 13, 19, and 23) to avoid oversampling the site. The quadrats were evenly spaced, approximately 15–30 m apart, depending on wetland area. Three circular quadrats were nested at each sampling point and all plants were identified to species as follows: herbaceous plants within 0.45 m² quadrats; shrubs within 25 m² quadrats; and trees within 100 m² quadrats. In addition, newly encountered plant species were recorded within a 2 m wide belt transect between each of the quadrats. The correlation between number of species sampled and the cumulative sampling area was analyzed at twelve study sites (Sites 2, 4, 5, 7, 9, 11, 12, 13, 21, 22, 23, and 27) by determining the plateau point of the cumulative-species-richness curve (after Mueller-Dombois and Ellenberg 1974) within the 2 m wide belt transect to assess the adequacy of the sampling effort. Among the sampled sites we estimated that the sampling effort was adequate; with approximately 85% of the plant species collected within the first half of the total transect length (an approximate cumulative sampling area of 400 m²).

Eight plant guilds were determined using fundamental plant structure types (per Reed 1988): (1) total plant community, (2) total woody plant community, (3) trees, (4) shrubs, (5) woody vines, (6) total herbaceous plant community, (7) emergent herbaceous plants, and (8) submersed herbaceous plants. Thirteen plant community parameters were tested in each guild among the

TABLE 3.2

Definitions of 13 Plant Community Parameters for 31 Depressional Wetlands in Central Ohio

Plant Guild Name	Plant Guild Code	Calculation	Source
Number of plant species	SPP	Count	Voss 1972, 1985, 1996
Number of plant genera	GEN	Count	Voss 1972, 1985, 1996
Number of plant taxonomic families	FAM	Count	Voss 1972, 1985, 1996
Percent of species on the Ohio Department of Natural Resources "invasive threat" plant list	%INVAS	Number of invasive species/SPP	Ohio Department of Natural Resources 1999
Percent of species with no wetland indicator status	%NI	Number of species with no wetland indicator status/SPP	Reed 1988
Percent of species with facultative-upland indicator status	%FACU	Number of species with facultative-upland indicator status/SPP	Reed 1988
Percent of species with facultative indicator status	%FAC	Number of species with facultative indicator status/SPP	Reed 1988
Percent of species with facultative-wetland indicator status	%FACW	Number of species with facultative-wetland indicator status/SPP	Reed 1988
Percent of species with obligate-wetland indicator status	%OBL	Number of species with obligate-wetland indicator status/SPP	Reed 1988
Percent of species within a given plant guild	%GLD	Number of species in each respective guild/ SPP	Reed 1988
Percent of species native to North America	%NNA	Number of native species native to North America/SPP	Andreas and Lichvar 1995
Perennial: annual ratio	PA	Number of perennial species/(Number of annual species + 1)	Reed 1988; Voss 1972, 1985, 1996
Percent of species that are primarily avian dispersed	%AD	Number of avian-dispersed species/SPP	Ridley 1930; Voss 1972, 1985, 1996

thirty-one wetlands (Table 3.2). Two traditional measures of information diversity applied to ecological communities (Magurran 1988) were tested in each guild, among the thirty-one wetlands. These diversity indexes were calculated for each plant guild at each wetland site: (1) Shannon's diversity index $(H') = -\Sigma p_i \ln p_i$ and (2) Simpson's index $= 1/(\Sigma p_i^2)$, where p_i is the proportion of individuals in the ith species. In this case study Simpson's index is

TABLE 3.3

Classes Used to Delineate Land Cover Surrounding 31
Depressional Wetlands

Land Cover Class	Regional Example(s)
Urban	Shopping center, parking area, suburban area
Agriculture	Corn or soybean row crop, grazed pasture
Grassland	Grassy "old field"
Forest	Forest patch
Open water	Pond, lake, stream
Bare soil	Gravel pit

Source: Anderson, J., E. Hardy, J. Roach, and R. Witmer. 1976. A land use classification system for use with remote-sensor data. U.S. Geological Survey Professional Paper 964. Washington, DC: U.S. Department of Interior.

expressed in the reciprocal form so that a large value indicates greater plant species heterogeneity.

Quantifying Land Cover in the Wetland Vicinity

A land cover analysis radius of 874 m was established, randomly selected from a criterion range of 500 to 1500 m, the approximate range of other studies of local land cover impact. The land cover analysis radius delineated the boundary of land cover *surrounding* the wetland (Table 3.1). Within the land cover analysis radius, contemporaneous aerial photographs (scale = 1:7,000) were used to map land cover classes in the vicinity of each wetland (Lyon 1981; Lillesand and Kiefer 1994; Jensen 1996; Lyon 2001). The land cover analysis radius and land cover polygons were manually traced onto a 21 cm × 28 cm acetate overlay using a 0.3 mm stylus. A dot-grid method was used to calculate the polygon area. Because of the relatively discreet land cover and land use types, relatively few classes (Table 3.3) and distinct class boundaries could be sufficiently delineated and calculated using manual polygon and dot-grid methods. Correlation analysis between dot-grid-derived wetland area and digitizer-derived wetland area ($\alpha = 0.77$, $P = 0.05$) indicated that these methods were sufficient to measure the land cover gradients among the thirty-one study sites.

Interwetland Distance Measurements and Statistical Analyses

The mean distance of neighboring wetlands was determined by measuring Euclidean distances to the three nearest wetlands at each study site on NWI maps ($N = 3$). Because one or both of the assumptions of parametric statistics tests (normality and equality of variance) are violated in all of the data, paired comparisons for all parameters were performed with the non-parametric

Spearman rank correlation (Zar 1984), α = 0.10. All statistical analyses were computed with Statview software (SAS Institute, 1998 v. 4.51).

Integrated Wetland Characterization and Assessment

Of the fifteen wetlands that were surrounded by less than the median areal percent of agriculture (i.e., <59% agriculture), seven have no agriculture surrounding them (Table 3.1). Of the other sixteen wetlands surrounded by agriculture (i.e., >59% agriculture), three sites were surrounded by a relatively large percent of urban land cover (Sites 13, 19, and 20). Twelve of these wetlands had a relatively large percent of forest land cover, mixed with the dominant agricultural land cover (Sites 1, 2, 5, 8, 9, 21, 22, 23, 24, 25, 26, and 27). The single remaining agriculturally dominated wetland (Site 3) was almost entirely surrounded by row-crop farmland (92.5%). The landscape surrounding four of the wetlands is predominantly urban land cover (Sites 12, 17, 18, and 28), with a median areal percent of 69%. These urban areas have a large proportion of impervious surfaces, likely reducing the amount of infiltration of water into the soil. Conversely, at Site 4 urban land cover is absent around the wetland. Median area of other land cover surrounding the wetland sites is 0.6% grassland, 19% forest, and 0.1% open water. Several wetlands had a relatively high percent of grassland (Sites 10, 14, 26, and 31), forest (Sites 4, 6, 10, and 31), open water (Sites 17, 18, 26, and 28), and bare soil (Sites 29 and 30). The relatively large percent of bare soil at Sites 29 and 30 is a result of a gravel pit approximately 800 m northwest of both wetlands, which was not hydrologically connected to either wetland. There were 399 plant species observed among the 31 wetlands, of which 79 are nonnative. Among the numerous correlations, the ecologically significant trends within specific plant guilds are detailed in Table 3.4 and Table 3.5. Taxa richness in at least one of the tested taxonomic groups (SPP, GEN, or FAM) is negatively correlated with interwetland distance for all guilds except submersed herbaceous plants (Table 3.4). There is a positive correlation between percent invasive plant species (%INVAS) and wetland area in the total herbaceous and submersed herbaceous plant guilds. Shannon's diversity index is negatively correlated with interwetland distance in the shrub, woody vine, and avian-dispersed plant guilds (Table 3.5). Simpson's index was negatively correlated with interwetland distance in the emergent herbaceous plant guild and in the total herbaceous plant community (Table 3.5). Thus, as interwetland distance increases, heterogeneity in the distribution of plant species at a wetland site decreases.

The taxa richness of submersed herbaceous plants (usually less than 20% of the plant species at a site) is not correlated with the interwetland distance, but is positively correlated with the area of open water in the local landscape ($P < 0.01$) and with the area of the wetland site itself ($0.01 < P < 0.05$). In the avian-dispersed plant guild, Simpson's index is negatively correlated with area of open water in the local landscape and with the area of

TABLE 3.4

Spearman Rank Correlation of Plant Guild Parameters and Land Cover Area in 8 Plant Guilds

	Area (ha)						
Parameter	Urban	Agriculture	Grassland	Forest	Open Water	Wetland Site	Interwetland Distance
Total Plant Community							
%INVAS	ns	ns	ns	ns	ns	ns	ns
SPP	ns	ns	ns	ns	ns	ns	ns
GEN	ns	ns	ns	ns	ns	ns	−0.516***
FAM	ns	ns	ns	ns	ns	ns	−0.453**
%NI	ns	ns	ns	ns	ns	ns	ns
%FACU	ns	ns	ns	ns	ns	ns	ns
%FAC	ns	ns	−0.309*	ns	ns	ns	ns
%FACW	ns	ns	0.302*	ns	ns	ns	ns
%OBL	ns	ns	ns	ns	0.384**	ns	ns
%GLD	0.500**	0.506**	0.546***	0.500**	0.557***	0.511***	0.500**
%NNA	ns	−0.309*	ns	0.587****	ns	ns	−0.324*
P:A	−0.360*	ns	ns	ns	ns	ns	−0.306*
%AD	ns	0.307*	−0.329*	ns	ns	ns	ns
Trees							
%INVAS	ns	ns	ns	ns	ns	ns	ns
SPP	ns	ns	ns	ns	ns	ns	ns
GEN	ns	ns	ns	ns	ns	ns	−0.355*
FAM	ns	ns	ns	ns	ns	ns	−0.377**
%NI	ns	ns	ns	ns	0.385**	0.415**	ns
%FACU	ns	ns	ns	ns	ns	ns	ns
%FAC	ns	ns	ns	ns	0.451**	ns	ns
%FACW	ns	ns	ns	ns	ns	ns	0.395**
%OBL	ns	ns	ns	0.306*	ns	ns	ns
%GLD	−0.311*	ns	−0.354*	ns	ns	ns	ns
%NNA	ns	−0.473**	ns	0.298*	ns	ns	ns
P:A	ns	ns	ns	ns	ns	ns	ns
%AD	ns	ns	ns	ns	ns	ns	−0.352*
Shrubs							
%INVAS	ns	ns	ns	ns	ns	ns	ns
SPP	ns	ns	ns	ns	ns	ns	−0.508***
GEN	ns	ns	ns	ns	ns	ns	−0.454**
FAM	−0.327*	ns	ns	ns	−0.321*	ns	−0.371**
%NI	ns	ns	ns	ns	ns	ns	ns
%FACU	ns	ns	ns	ns	−0.528***	ns	ns
%FAC	ns	ns	ns	−0.323*	ns	ns	ns

(continued)

TABLE 3.4

Spearman Rank Correlation of Plant Guild Parameters and Land Cover Area in 8 Plant Guilds (continued)

	Area (ha)						
Parameter	Urban	Agriculture	Grassland	Forest	Open Water	Wetland Site	Interwetland Distance
%FACW	ns	ns	ns	ns	ns	ns	ns
%OBL	−0.523***	0.312*	ns	ns	−0.320*	ns	−0.445**
%GLD	−0.323*	0.343*	ns	ns	−0.329*	−0.315*	ns
%NNA	ns	ns	ns	ns	ns	ns	ns
P:A	ns	ns	ns	ns	ns	ns	−0.395**
%AD	ns	ns	ns	ns	ns	ns	ns
Woody Vines							
%INVAS	ns	ns	ns	ns	ns	ns	ns
SPP	ns	ns	ns	ns	−0.301*	ns	−0.542***
GEN	ns	ns	ns	ns	−0.299*	ns	−0.477**
FAM	ns	ns	ns	ns	ns	ns	−0.372**
%NI	i.d	i.d	i.d	i.d	i.d	i.d	i.d
%FACU	ns	ns	ns	ns	ns	ns	−0.579****
%FAC	ns	ns	ns	ns	ns	ns	ns
%FACW	na	na	na	na	na	na	na
%OBL	na	na	na	na	na	na	na
%GLD	−0.328*	ns	ns	ns	−0.416**	ns	−0.344*
%NNA	−0.334*	ns	ns	ns	−0.480**	−0.344*	ns
P:A	ns	ns	ns	ns	-0.380**	ns	−0.371**
%AD	−0.434**	0.353*	ns	ns	−0.485**	−0.421**	ns
Total Woody Plants							
%INVAS	ns	ns	ns	ns	ns	ns	ns
SPP	−0.322*	ns	ns	ns	ns	ns	−0.462**
GEN	−0.301*	ns	ns	ns	ns	ns	−0.469**
FAM	−0.317*	ns	ns	ns	ns	ns	−0.402**
%NI	ns	ns	ns	ns	0.403**	ns	ns
%FACU	ns	ns	ns	ns	−0.438**	ns	−0.439**
%FAC	ns	ns	ns	ns	0.409**	ns	ns
%FACW	ns	ns	ns	ns	ns	ns	0.407**
%OBL	−0.504**	ns	ns	ns	ns	ns	−0.483**
%GLD	−0.362*	0.336*	ns	ns	−0.364*	ns	ns
%NNA	ns	−0.527***	ns	ns	ns	ns	ns
P:A	ns	ns	ns	ns	ns	ns	ns
%AD	ns	ns	ns	ns	ns	ns	−0.491**
Emergent Herbaceous Plants							
%INVAS	ns	ns	ns	−0.404**	ns	0.352*	ns

(continued)

TABLE 3.4

Spearman Rank Correlation of Plant Guild Parameters and Land Cover Area in 8 Plant Guilds (continued)

Parameter	Urban	Agriculture	Grassland	Forest	Open Water	Wetland Site	Interwetland Distance
SPP	ns	ns	0.382**	ns	ns	ns	−0.322*
GEN	ns	ns	0.384**	ns	ns	ns	−0.371**
FAM	ns	ns	0.400**	ns	ns	0.369**	ns
%NI	ns	ns	ns	ns	−0.370**	ns	ns
%FACU	ns	0.339*	−0.374**	−0.331*	−0.390**	−0.320*	ns
%FAC	ns	ns	ns	ns	ns	ns	ns
%FACW	ns	ns	0.356*	ns	ns	ns	ns
%OBL	ns	ns	ns	ns	0.412**	ns	ns
%GLD	0.451**	−0.470**	0.359*	ns	0.311*	ns	ns
%NNA	ns	−0.344*	ns	0.473**	ns	ns	−0.318*
P:A	ns	ns	ns	ns	ns	ns	ns
%AD	ns	ns	ns	−0.307*	ns	ns	0.412**
Submersed Herbaceous Plants							
%INVAS	ns	ns	ns	ns	0.331*	0.303*	ns
SPP	0.343*	ns	ns	ns	0.622****	0.466**	ns
GEN	0.363*	ns	ns	ns	0.530***	0.538***	ns
FAM	0.364*	ns	ns	ns	0.523***	0.519***	ns
%NI	ns	ns	ns	ns	ns	ns	ns
%FACU	ns	ns	ns	ns	ns	ns	ns
%FAC	ns	ns	ns	ns	ns	ns	ns
%FACW	ns	ns	ns	ns	ns	ns	ns
%OBL	0.421**	ns	0.307*	ns	0.655****	0.458**	ns
%GLD	0.363*	ns	ns	ns	0.663****	0.444**	ns
%NNA	0.414**	ns	ns	ns	0.651****	0.388**	ns
P:A	0.336*	ns	ns	ns	0.632****	0.452**	ns
%AD	0.402**	ns	0.307*	ns	0.648****	0.420**	ns
Total Herbaceous Plants							
%INVAS	ns	ns	ns	ns	ns	0.421**	ns
SPP	ns	ns	0.401**	ns	ns	ns	ns
GEN	ns	ns	0.384**	ns	ns	ns	−0.371**
FAM	ns	ns	0.401**	ns	ns	0.360*	ns
%NI	ns	ns	ns	ns	−0.376**	ns	ns
%FACU	ns	ns	ns	ns	−0.467**	ns	ns
%FAC	ns	ns	ns	ns	ns	ns	ns
%FACW	ns	ns	0.440**	ns	ns	ns	ns
%OBL	ns	−0.319*	ns	ns	0.363*	ns	ns

(continued)

TABLE 3.4

Spearman Rank Correlation of Plant Guild Parameters and Land Cover Area in 8
Plant Guilds (continued)

	Area (ha)						
Parameter	Urban	Agriculture	Grassland	Forest	Open Water	Wetland Site	Interwetland Distance
%GLD	0.351*	−0.360*	ns	ns	0.327*	ns	ns
%NNA	ns	−0.401**	0.326*	0.514***	ns	ns	ns
P:A	ns	ns	ns	ns	ns	ns	ns
%AD	ns	ns	ns	ns	ns	ns	0.425**

Notes: Correlation (Rho) values and significance are shown (N = 31); ns = not significant; * = 0.10,
** = 0.05, *** = 0.001; id = insufficient data; na = parameter is not applicable.

the wetland (Table 3.5). Highly significant positive correlations exist between
the area of open water in the vicinity of the wetland and the taxa richness,
perennial:annual ratio, percent native plant species, percent obligate wetland
plant species, percent avian-dispersed plant species, and guild percent (i.e.,
%GLD) in the submersed herbaceous plant guild (Table 3.4). The same trend,
albeit less significant, exists for percent urban land cover. There was a nega
tive correlation between Shannon's diversity index and percent urban land
cover in the shrub guild and a similar (yet weaker) relationship in the total
woody plant community (Table 3.5). Taxa richness values were not correlated with wetland area, but they were weakly (positively) correlated in the
total plant community and the total woody plant community. In the submersed herbaceous plant guild, taxa richness parameters were positively
correlated with wetland area, area of urban land cover, and area of open
water (Table 3.4). Exceptionally strong correlations exist between: (1) percent
native plant species and the area of forest surrounding a wetland (α = 0.587,
$P < 0.001$) in the total plant community and (2) the percent facultative-upland
plant species and interwetland distance (α = −0.579, $P < 0.001$) in the woody-
vine guild.

Landscape-Scale Wetland Characterization Lessons Learned

In general, the distance a disseminule has to travel can be an important factor in the dispersal of plant species between wetlands (Van der Valk 1981),
and the results demonstrate how the combined effects of wetland area and
interwetland distance may affect the dominance of herbaceous plant species in depressional wetlands. The abundance of woody-vine species may
be a combined result of close proximity of neighboring wetlands, greater
vertical structure of wooded sites and the relative absence of human use
in the surrounding landscape that facilitates interwetland dispersal of
woody-vine plant species. An increase in the presence of grassland or forest
in the surrounding landscape has been demonstrated to facilitate the avian

TABLE 3.5

Spearman Rank Correlation of Land Cover Area, Wetland Site Area, and Interwetland Distance with Diversity Indexes among 8 Plant Guilds

| Plant Guild | Index | Area (ha | | | | | Wetland Site | Interwetland Distance |
		Urban	Agriculture	Grassland	Forest	Open Water		
Total plant community	Shannon's	ns	ns	ns	ns	ns	ns	ns
Total plant community	Simpson's	ns	ns	ns	ns	ns	ns	ns
Trees	Shannon's	ns	ns	ns	ns	ns	ns	ns
Trees	Simpson's	ns	ns	ns	ns	ns	ns	ns
Shrubs	Shannon's	−0.488**	ns	ns	ns	ns	ns	−0.545**
Shrubs	Simpson's	0.555*	ns	ns	ns	ns	ns	ns
Woody vines	Shannon's	ns	−0.436*	ns	0.433*	ns	ns	−0.535**
Woody vines	Simpson's	ns	ns	ns	ns	ns	ns	ns
Total woody plants	Shannon's	−0.440*	ns	ns	ns	ns	ns	ns
Total woody plants	Simpson's	ns	ns	ns	ns	ns	ns	ns
Emergent herbaceous plants	Shannon's	ns	ns	ns	ns	ns	ns	ns
Emergent herbaceous plants	Simpson's	ns	ns	ns	ns	ns	ns	−0.486**
Submersed herbaceous plants	Shannon's	ns	ns	0.404*	ns	ns	0.331*	ns
Submersed herbaceous plants	Simpson's	ns	id	id	id	id	id	ns
Total herbaceous plants	Shannon's	ns	ns	ns	ns	ns	ns	ns
Total herbaceous plants	Simpson's	ns	ns	ns	ns	ns	ns	−0.395*
Total avian-dispersed plants	Shannon's	ns	ns	ns	ns	ns	ns	−0.537**
Total avian-dispersed plants	Simpson's	ns	ns	ns	ns	−0.590***	−0.523**	ns

Notes: Correlation (Rho) values and significance are shown ($N = 21$); ns = not significant; * = 0.10, ** = 0.05, *** = 0.01; id = insufficient data; na = parameter is not applicable.

dispersal of plant species between neighboring ecosystems (McDonnell and Stiles 1983; McDonnell 1984; Dzwonko and Loster 1988), implicating birds as one possible overcomers of distance between depressional wetlands. As reported in upland forests of Central Ohio (Simpson et al. 1994; Stritthold and Boerner 1995), a decline in the biological diversity of Ohio's depressional wetlands may be a result of the predominant agricultural landscape that surrounds them. For example, Shannon's diversity is negatively correlated with interwetland distance for shrubs, woody vines, and avian-dispersed plants. Heterogeneity of species distribution (i.e., Simpson's index) in the total plant community is also negatively correlated with interwetland distance. Thus in landscapes where wetlands are isolated from each other (e.g., by agriculture) the total herbaceous plant community is likely to be less diverse and more homogeneous than where wetlands are in close proximity to each other. In Central Ohio, such human-induced fragmentation of forests is much more extensive than in other areas of the country (Dahl 1990). Thus it may be that human-induced landscape fragmentation in Central Ohio has increased the predominance of disturbance-adapted plant species in the remnant patches of depressional wetlands.

None of the woody plant guilds are significantly correlated with area of grassland or forest with regard to taxa richness, suggesting that the woody plants may not be as sensitive an indicator of relatively *natural* landscapes as are the herbaceous plant guilds. This is probably a result of the fact that woody plants grow slower than herbaceous plants and direct much of their photosynthate toward the production of structural plant tissue for long-term survival (Chapin 1991). Comparatively, herbaceous plants grow (and senesce) quickly and direct much of their photosynthate toward reproduction and short-term survival. Thus, woody plants have longer response times than herbaceous plants because they are relics. Accordingly, the woody plants contain greater *ecological inertia* than the herbaceous plant guilds. The reason why taxa richness in the total plant community is not similarly correlated with the area of grassland or forest may be a result of the ecological inertia of the woody plants (i.e., plant guilds that comprise longer lived species that rarely invade), which tempers the ecological momentum of the herbaceous plant guilds. Thus invasions of wetlands by incoming plant disseminules would be likely to proceed in a shorter period of time for herbaceous plant species than for woody plant species because of longer-term recruitment in woody plants, but sudden wetland disturbances may cause a rapid (negative) change in the herbaceous plant community.

The influence of wetland size, open water, and interwetland distance on the floristic characteristics of depressional wetlands may be a result of dispersal constraints for plants among depressional wetlands. For example, dispersal may be limited by the smaller target area that a depressional wetland provides for incoming disseminules, thereby negatively affecting the likelihood that disseminules will arrive in smaller wetlands. Thus the number of plant species in depressional wetlands generally decreases for smaller wetlands and increases for larger wetlands. For avian-dispersed plant species, the dominance of certain plants may increase with area. The correlations observed between landscape characteristics and wetland plant guild parameters in this case study would not have been detected if we had solely compared the mean differences among sites. This is one of the advantages of using gradient analyses, because it allows for the detection of a finer degree of differences between individual wetland site characteristics. But testing the differences between extreme parameter values and mean parameter values could be useful if large differences existed between many of the wetlands or if establishing parameter groups a priori were desired prior to gradient analyses. In this case study gradient analyses, along with the use of plant guilds, allowed for the fullest exploration of the ecological variability among wetlands as possible (Lyon and Lyon 2011).

Correlations between the wetland plant community parameters and the surrounding land cover characteristics, interwetland distance, and wetland area suggest that wetland plant guilds may be a commonsense biological assessment tool to detect impacts on wetlands (longitudinal studies) or for detecting impacts among a group of wetlands in varied landscape conditions (cross-sectional studies).

Studies such as this one may be useful for quantifying the relative effects of landscape conditions on riverine and other natural wetlands, or to track wetland restoration projects over time. From such studies, the relationships between the wetland plant community parameters and landscape stressors may become better understood among a variety of different wetland ecosystem types, at which time they could be used to predict general shifts in the condition of ecologically vulnerable wetlands.

Broad Assessment of Rivers, Stream, and Associated Wetlands

As suggested by the work in geographically isolated wetlands, landscape characterization can allow for certain, specific evaluations of high-valued riverine wetlands too. Potentially, remote sensor measurements can help to identify areas of undermining of river and stream crossings. Remote sensors can also identify the extent of deposition and erosion of floodplain sediment materials, as well as periodic connections with geographically isolated wetlands, and other ecosystems or features in the landscape. These indicators can help identify areas for checking by field personnel, and potential avoidance of the failure of structures or associated impacts involving floodplain wetlands and riparian areas.

Wetland Ecosystem and Receptor Monitoring

The issue of wetland resources monitoring is important for tracking the current, historical, and future quantities and varieties of wetlands. These efforts may range from straightforward assessments of quantity and change over time, or they may be complex evaluations of quantity and variety of wetlands as influenced by stressors.

To address the variety of information needs related to planning and management of wetlands, aquatic and terrestrial ecosystems often require analyses of data over time or monitoring. A monitoring approach should incorporate capabilities to collect and compare data. Often these methodologies are termed *change detection* methods and measurements (Lunetta and Elvidge 1998; Lyon et al. 1998).

There are a number of ways to monitor wetland, aquatic, and terrestrial ecosystems over time. Each has a value, and each may be used for certain requirements. Methods that involve imaging measurements have been found useful in these applications. These imaging and other methods include field inspection and monitoring, aerial photograph use and interpretation, aerial or ground videography, and aircraft and satellite data processing.

A change detection or monitoring methodology can be greatly facilitated by the use of digital imaging data from sensors. Imaging data have the capabilities to supply spatial and spectral data over large areas, and do so at a potentially lower relative cost than fieldwork and mapping.

Methods for Characterizing Changes in Features

The change in land cover acts as an indicator of activities. Methodologies for detection of change have been examined over a period of time. Remote-sensor-based technologies offer a great deal of capability for wetland and wetland-related applications.

Indicators of construction are an important identifier of activity, and help to locate areas of new housing or buildings. The new construction areas can be identified by the removal of vegetation and presence of bare earth. The bare earth is commonly made up of mineral soil, which has a high reflectance of light, and may indicate disturbances. Bare mineral earth appears very light toned on black-and-white film or bright white-brown on color imagery. Similarly, and on occasion, soils high in organic matter are disturbed. These high organic matter soil areas appear to be very dark toned and darker than plants in the visible part of the spectrum. Monitoring in such areas of disturbance can demonstrate construction activities by the changes that occur, from a vegetated or other land cover compliment to the lighter-toned (bare earth) areas of construction. This difference is very simple to detect from satellite or aerial data.

This monitoring and indicator method can identify new construction for a number of years after the fact, and it takes time for native or invasive vegetation to repopulate the disturbed soils or for landscaped plants to mature. Development of structures results in bare earth areas, and subsequent growth of grass, shrubs, and trees (which are much different from the original land cover) are determinable by comparison of pattern and tone or color on imagery.

To meet the traditional and new requirements for information in support of government, industry, and conservation and environmental groups, digital technologies are utilized that supply appropriate data. Digital technologies must be flexible to address both traditional and new problems and questions. Digital imaging methods have the necessary capabilities to supply the appropriate data for broad-scale analyses. The appropriate data for monitoring must also be available at a relatively low cost. In a monitoring approach, there is the implicit need to acquire image data from two or more periods of time, which are available frequently in the United States from the Landsat Program and other aerial data acquisition programs.

Digital remote sensor technologies provide appropriate methods for monitoring the characteristics of rivers, wetlands, and adjacent land cover types. Implementation of a monitoring approach requires a combination of input data sets, as well as hardware and software for digital data processing. The effort is straight forward, and can be accomplished by employing technicians or other experts.

Most problems and questions can be addressed using a combination of digital remote sensing and GIS technologies. The resulting data products can be used to determine the presence or absence of existing features. The capabilities of digital remote sensing and GIS allow for analyses of larger areas and many data sets.

Case Study: Landsat Characterization of Map Analysis of Riparian and Riverine Wetlands of the Mississippi River (Headwaters to the Gulf of Mexico)

The previous case study demonstrated several detailed uses of imagery for assessing wetlands at a landscape scale using traditional (i.e., nondigital) methods for identification and interpretation techniques, which is critical for comprehending the bases of contemporary (i.e., digital) techniques. Analogously, satellite data can be used to provide contextual information across more extensive areas of the landscape due to their (typically) broader field of view.

The purpose of this case study was to provide a foundational reference for ecological vulnerability throughout the entire Mississippi River Basin (see Appendix) using publically available digital (Landsat) data. The case study's resultant map series provides the reader with an easy reference for further review, and includes basic reference information so that additional ecological vulnerability analyses could be conducted using existing water quality data, existing hydrologic data, and land cover data in targeted finer-scale areas, where desired.

Characterization of Vegetation

An advantage of using color images, from the near-infrared part of the spectrum (0.7–1.1 μm) is that healthy, vigorously growing plants reflect a great deal of near-infrared light. These plants or vegetation appear bright white-toned on black-and-white infrared photos, or bright pink or magenta on color infrared film or sensors. These characteristics allow the user to identify live plants from dead plants or plant residue. The user can identify wetland plants growing at or above the surface of the water (floating leaved and emergent wetland plants, respectively) interpreted using near-infrared imagery, as previously described.

Conversely, plants growing beneath the water surface, or submergent wetland plants or submergent (also known as *submersed*) aquatic plants, do not exhibit the bright tone or pink color of emergent plants, because the near-infrared light is largely absorbed by water and not reflected back to the sensor. This difference allows the user to identify emergent wetland plants and terrestrial plants, and distinguish them from submergent wetland or submergent aquatic plants, which tend to appear dark toned or dark blue in color on black-and-white or color infrared film or sensor products (Lyon and Olson 1983; Lyon 1993), respectively.

As addressed in previous sections, the appropriate methods for characterizing vegetation in wetlands are based on the project goals, and the scale or resolution of the application. Depending on scale, vegetation communities can be identified by plant species in the field. The characteristics of the vegetation communities can be identified from a distance by their shape, size, tone, or color and other diagnostic characteristics.

Soil Characterizations

Photointerpretation of aerial photos and remote sensing experiments have been performed over the years with an eye toward characterizing soil features and moisture conditions. Important advances have been made by using the middle-infrared portion of the spectrum, as it is sensitive to moisture conditions of soils and plants. Good work has also been done using multiple dates of Landsat Thematic Mapper middle-infrared data (bands 5 and 7) to identify hydric soil conditions. Lunetta et al. (1999) found that the combination of spring, leaf-off imagery, and growing season imagery allowed for increased detection of hydric soil conditions that are often obscured by leaf canopy cover in the growing season.

Remote sensing data can also be used to help distinguish between mineral soils and soil high in organic matter. The former will have a different color or spectra in the red and green regions of the spectrum, as compared to the dark tone or spectra of the very organic soils.

Water Resource Characterization

Remote sensing data and field-based data are particularly good for identifying hydrologic characteristics of wetlands, which can change rapidly over time. Remote sensing data records the reflected or upwelling light from water and the result is good detail from water or aquatic ecosystems as distinct areas from terrestrial ecosystems. The near-infrared portion of the spectrum has been used to identify plant and hydrologic conditions of wetlands. This is due to the unique light or spectral characteristics of water and plants on color infrared or black-and-white infrared film, or from instruments measuring light in the near-infrared portion of the spectrum.

As described earlier, the majority of near-infrared energy is absorbed by water. A very small portion is reflected or transmitted through water. The net result is that water with very low concentrations of suspended sediments or chlorophyll in phytoplankton will absorb most infrared light. Recall that this causes water to appear dark toned or dark black-blue on black-and-white and color infrared films, respectively (Lyon and Olson 1983; Lyon 1993; Ward and Elliot 1995).

The behavior of near-infrared energy and water facilitates the interpretation of certain conditions of wetlands, lakes, and all coastal areas. The first advantage of near-infrared film or near-infrared sensor instrument measurements is the detection of the edge of water and soil. Near-infrared light reveals the edge between water and shore because very small thicknesses of water will absorb near-infrared energy, and the edge is identified clearly. Hence, infrared film or sensors can identify the edge of water bodies. This interaction is particularly stark in arid areas that have some surface water during limited portions of the year (e.g., desert springs, oases, or marshes).

Conversely, film or sensors working in the visible part of the spectrum experience penetration of light through the water to the bottom and reflectance to the surface (Lyon et al. 1992; Lyon and Hutchinson 1995).

Wetland Classification

In inventory and mapping of wetlands and other land cover types, it is necessary to identify the wetlands as to category or type. Often this activity is called thematic mapping (i.e., identifying land cover *themes*). This is similar to the mapping of geological types or themes, or mapping of land ownership.

There are a number of wetlands types and classification systems to describe them (Anderson et al. 1976; Cowardin et al. 1979; U.S. Army Corps of Engineers 1987; U.S. Environmental Protection Agency 1991; Brinson 1993). These types and classification systems are associated with the hydrological, soil or edaphic characteristics, and vegetation characteristics of wetlands. Other systems that characterize wetlands are associated with the functions and uses of wetlands (Adamus et al. 1991; Brinson 1993). A commonly used system in the United States is the Cowardin System of the U.S. Fish and Wildlife Service (USFWS; Cowardin et al. 1979). It is employed in describing wetland areas by vegetation and hydrologic regime types, allowing for the mapping of wetland types of the United States in the NWI and a series of map/data products.

Case Study: Digital Characterization Approaches for Geographically Isolated Wetlands

The following case study integrates digital and traditional photointerpretive (i.e., traditional) analyses to characterize the geographically isolated wetlands (Leibowitz and Nadeau 2003) of southeastern Texas, and describes a cost-effective and accurate method for using both airborne and satellite remote sensing to monitor wetlands at multiple scales to prioritize finer-scale assessment and restoration work. The results described in this case study are based on maps of palustrine wetlands (Cowardin et al. 1979), addressing national legislation and several important Supreme Court rulings, and associated appeals and litigation (e.g., *Rapanos v. United States of America* and *Carabell v. U.S. Army Corps of Engineers*).

A major portion of effective wetland mapping is the detailed exploration of map accuracy. This case study builds on the traditional photointerpretive approach in Ohio, earlier in this chapter, by presenting map information in the context of its accuracy so that the users of these data can be aware of the advantages and limitations of using the characterization results. The methodology used to produce the accuracy assessment is presented in detail so that map results can be placed in the context of the uncertainty that is always a factor in all remote sensing projects (Foody and Atkinson 2002).

It is important to recognize that palustrine wetlands, and subclasses thereof (Cowardin et al. 1979), are among the most diverse and ephemeral of ecosystems, which poses unique challenges for the effective use of remote sensing data for wetland detection and mapping. Additional options for data use include products available from the Federal Emergency Management Agency (FEMA), which utilize Light Detection and Ranging (LIDAR) data to precisely map terrain, wetlands, and flood risks.

The accuracy assessed Landsat-based maps produced by the techniques presented in this case study have been used to quantify the total number and total area of palustrine wetlands within the 100-year floodplain boundaries and within the 100- to 500-year floodplain boundaries per current FEMA Q3 Flood Data (FEMA 1996). The total number and total area of palustrine wetlands were also calculated for areas outside of the FEMA flood zones (i.e., outside of the 100-year and the 100- to 500-year floodplain zones). All floodplain calculations have been initially presented in this case study by county. The purpose of the tabulated data is to provide an initial estimate of the relative presence of open water, emergent, scrub-shrub, and forested wetlands (by count and by area) in floodplain zones and the relative contribution of palustrine wetlands among the five counties in the vicinity of Houston, Texas.

Note that any wetland that falls among two or more floodplain zones (e.g., part of a wetland is in the 100-year floodplain and another part of the same wetland in the 100- to 500-year floodplain) was included in both floodplain zones by count and by total wetland area. This provides an assessment of wetlands in each of the floodplain zones, however overestimates the total number and total wetland area if the individual county values are summed. Therefore, summing the values in this case study among different counties will result in an overestimation of wetland count and area values, and is not recommended.

Wetland data were thematically mapped for the coastal Texas study area through the classification of four multitemporal sets (spring, fall, early winter, and late winter) of Landsat Enhanced Thematic Mapper (ETM+) data. Four scenes of the ETM+ satellite data cover the five counties of Harris, Fort Bend, Brazoria, Montgomery, and Galveston that comprise the study area (Figure 3.10). The satellite data were selected from available coverage with minimum cloud and haze conditions. The collection of all imagery in a single

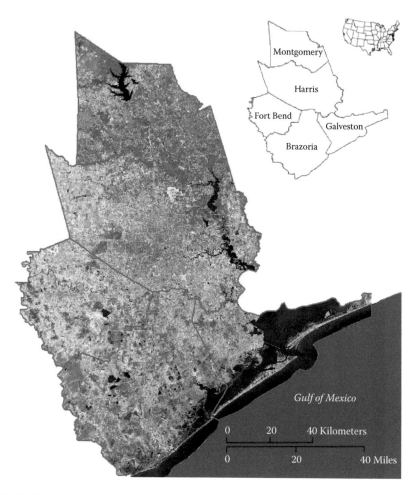

FIGURE 3.10
(See color insert.) The study area is the combined area of Montgomery County, Harris County, Fort Bend County, Brazoria County, and Galveston County, Texas. The remote sensing data shown is generally centered on the metropolitan area of Houston, Texas.

recent year was preferred but was not possible because of excessive cloud cover or haze. These and other image quality factors (such as image striping and banding) required the selection of satellite data spanning the years of 1999 to 2003. Matched cloud-free scene dates were collected. Summer imagery would have been ideal, allowing for analyses based on all four seasons, but almost all of the available summer imagery in the study area contained substantial cloud cover (by area) and thus violated our acceptance criteria and were not used. Imagery from 2001 was avoided due to the drought conditions during that year. The satellite data used to produce the land cover map, from which wetlands for this case study were mapped, is comprised of

TABLE 3.6

Landsat Satellite Scenes Used for Analysis, which Met Acceptance Criteria

Path/Row	Spring Date	Fall Date	Early Winter Date	Late Winter Date
25/39	30-Mar-00	6-Oct-99	9-Nov-00	15-Jan-02
25/40	23-Mar-03	6-Oct-99	9-Nov-00	15-Jan-02
26/39	30-Mar-03	29-Sep-00	6-Nov-02	23-Feb-02
26/40	30-Mar-03	29-Sep-00	6-Nov-02	23-Feb-02

four satellite scenes (i.e., tiles of data) acquired during the spring, fall, early winter, and late winter months (Table 3.6).

The spatial resolution of the Landsat ETM+ satellite data were nominally 30 × 30 m (i.e., a total nominal pixel area of 900 m^2). Thus objects and materials on the ground smaller than 900 m^2 may be confused for other objects and materials. For example, a 10 × 10 m patch of emergent wetland vegetation directly adjacent to a 10 × 10 m patch of water may be erroneously mapped as water or some other land cover type with similar reflectance characteristics to the apparent mixture of water and vegetation. The accuracy of the classification of satellite data may also be affected by the alignment of the patches (on the ground) with the pixel alignment; the configuration, orientation, and shape of adjacent patches; and the interspersion and mixture of materials that fall within the 30 × 30 m pixel location (e.g., other vegetation and soil underlying the vegetation). Despite the 30 × 30 m pixel accuracy constraints, the advantages of using Landsat data are several-fold: (1) large areas of the landscape may be repeatedly mapped at a relatively low cost (e.g., compared to traditional aerial photograph manual interpretation costs); (2) seasonal conditions may be monitored to assess the hydroperiod of wetlands; and (3) historical change analyses may be conducted.

Aerial photographs were used as a source of high-resolution reference data for conducting the accuracy assessment of the land cover map produced by classification of satellite data. To control for landscape change over time it was preferable to obtain stereo aerial photography and photography dates as close to the satellite data acquisition date as possible. For this case study, a set of eighteen stereopairs of 1995 National Aerial Photography Program (NAPP) color infrared imagery was used to produce the reference data for the accuracy assessment of the land cover map (Figure 3.11). The eighteen stereopairs, each covering 3.2 km^2 areas, provided a sample of the study area and were visually selected to provide an even distribution of samples in areas where wetlands predominate, that is, not in the urban center of Houston. A five-class wetland map for the coastal Texas study area was generated using the multiple-date, multiple-season set of Landsat Thematic data and existing NWI maps. All the data used to produce the map were coregistered with each other. An accuracy assessment, based upon two sets of error

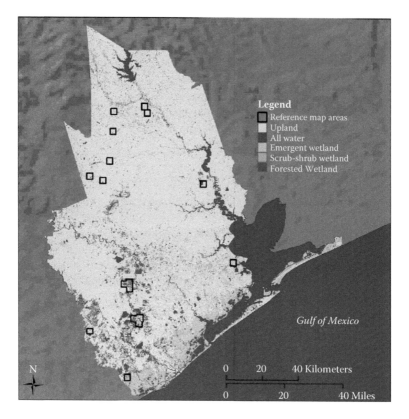

FIGURE 3.11
(See color insert.) Eighteen stereopairs of aerial photographs (shown as individual or clustered sample areas) were used to check the accuracy of the satellite-based classification with the apparent presence of wetlands on the ground.

matrices and associated statistical results, was performed using 1995 aerial photographs as a high-resolution reference source.

The study area was the surrounding area of Houston, Texas, delineated by the combined areas of Harris County, Fort Bend County, Brazoria County, Montgomery County, and Galveston County, Texas (Figure 3.10). Total area of the five-county study area was 16,086 km², which contained a diverse landscape containing urban, agricultural, and natural land cover types (Table 3.7). The study area was located within three different Level III Ecoregions (Omernik 1987): (1) the Western Gulf Coastal Plain, (2) the South Central Plains, and (3) the East Central Texas Plains. This region of Texas receives an average of 122 cm of precipitation per annum and has an average annual temperature range from 41°F to 94°F (5°C to 34°C).

TABLE 3.7

1990s National Land Cover Data (NLCD) Set Information for the 5 Counties

Land Cover Type	Brazoria County	Fort Bend County	Galveston County	Harris County	Montgomery County
Open Water	18968	4740	10963	13467	9063
Low intensity residential	6766	5125	9051	67911	10234
High intensity residential	3273	4019	4785	50117	3317
High intensity commercial/ industrial/transportation	5706	3669	5953	45975	4716
Bare rock/sand/clay	572	939	406	2386	812
Quarries/strip mines/gravel pits	86	271	140	225	404
Transitional	0	90	0	35	3583
Deciduous forest	46231	31145	10156	30772	19815
Evergreen forest	60725	11223	13636	17451	50290
Mixed forest	137	261	0	56168	113835
Deciduous shrubland	6310	8171	939	4626	0
Planted/cultivated (orchards, vineyards, groves)	2	2	0	0	0
Grassland/herbaceous	15629	8706	6233	3417	1
Pasture/hay	94291	100412	9820	115798	53127
Row crops	25792	32549	1360	1303	1146
Small grain crops	6219	1976	569	5202	0
Other grasses (urban/rec; e.g., parks, golf courses)	1101	1948	951	24162	1704
Woody wetlands	14983	6806	711	2091	7216
Emergent herbaceous wetlands	64273	7566	30395	6512	3355

Image Processing

Image processing techniques used to generate the wetlands land cover data for the coastal Texas study area included image preparation and image classification. The Landsat ETM+ satellite data were normalized, georegistered, and mosaicked together to simplify processing. Imagine (ERDAS, v. 8.7) image-processing software was used for satellite data subsampling, classification, and output.

Atmospheric Correction

Atmospheric correction methods are often used for multidate image analysis. The objective is to reduce atmospheric variations between various dates of imagery so that comparisons are made on *pure* reflectance properties of objects under study. To better normalize the data values and minimize extraneous variations and improve signature extension across large areas, the data were processed to minimize atmospheric effects and reduce the dimensionality of the data.

Dark Object Subtraction

One of the simplest methods of atmospheric correction is the dark object subtraction method. The assumptions are that there is uniformity in the atmosphere throughout the entire image and that a true dark object exists in the image. A true dark object is one that absorbs energy, such that energy from the object is 0. If the assumptions are true, any energy returned to the sensor from the dark object is most likely caused by atmospheric scattering. Therefore the returned value can then be subtracted from each pixel of the image to provide an atmospherically corrected image. The following formula was applied: Ls = Lt + Lh, where Ls is the energy reflected and emitted from the ground (Lt) and Lh is the energy from scattering. Therefore any return from the dark body is likely from scattering. The following formula was modified for a dark object: Lsdo = 0 + Lh, where Lsdo is the energy reflected from a dark object, plus scattered energy (Chavez 1988). An Imagine-based model was developed to automate the dark object method. The model assumes that the lowest value in each image band represents a dark object. It does not account for anomalies in the data set, and it ignores the value zero.

Radiance to Reflectance and Conversion

To reduce the variation in atmospheric conditions, image data values can be converted from radiance to reflectance. Radiance is the amount of radiation from an object, and it is influenced by both the properties of the object and the sun energy hitting the object. In the case of earth imaging, radiance is therefore not always a good indicator of physical properties. Radiance will vary with the time of day and atmospheric conditions. Converting Landsat data into reflectance can provide more useful information about the properties of objects. Imagine software contains a prebuilt tool for converting Landsat radiance to reflectance based on seasonal and time-of-day sun angle factors. Once the Landsat data were converted to reflectance values, the Kauth and Thomas (1976) approach was applied to process the multiple bands of Landsat data into normalized brightness, greenness, and wetness indicators. This was accomplished using coefficients derived from the Landsat 7 ETM+ sensor (Huang et al. 2001).

Image Registration

High-resolution image registration, with a minimum spatial offset, was a critical part of the multitemporal (i.e., multiyear, multiseason) wetland detection approach used in this research. Thus image registration acceptance criteria were set at a root mean square error (RMSE) of less than 1 pixel. During these steps, we determined that a second order transformation of the data resulted in the best image-to-image registration.

Wetland Classification

Initial classification of the Landsat data identified all wetland areas. Subsequent subsetting of the wetland data for the thematic mapping effort was attempted to distinguish solely palustrine wetland classes (from Cowardin et al. 1979), summarized in Figure 3.12:

Open water palustrine wetlands—Areas that may have fluctuating hydrology that may be observed at some time of the year as open water, or water with floating or submersed vegetation. Classes included in this type of wetland are rock bottom, unconsolidated bottom, aquatic bed, rocky shore, and unconsolidated shore.

Emergent palustrine wetlands—Characterized by erect, rooted, herbaceous hydrophytes, excluding mosses and lichens. This vegetation is present for most of the growing season in most years. Perennial plants usually dominate these wetlands.

Scrub-shrub palustrine wetlands—Includes areas dominated by woody vegetation less than 6 m (20 ft) tall. The species include true shrubs, young trees, and trees or shrubs that are small or stunted because of environmental conditions.

Forested palustrine wetlands—Characterized by woody vegetation that is 6 m tall or taller.

Modeling Techniques and Indicators

The primary training data for this project were the existing (circa 1980s) National Wetland Inventory (U.S. Fish and Wildlife Service, various dates). Initially, approximately 100,000 points were randomly located within the training sites using the Random Points extension for ArcView 9 (Environmental Systems Research Institute, v. 3.3). The random points were converted to ARCINFO grids and then to Imagine (.img) files. Each Imagine pixel was attributed with the appropriate wetland code of interest. The data set was then used for production of the wetland map using the classification and regression tree (CART) modeling process.

Sample pixels were overlaid on each of the predictor data layers to produce a data set containing both predictor (imagery) variables and the response variable (wetlands label code) using the CART Sampling Tool of the CART Module (EarthSatellite Corporation 2003). See5 data mining software

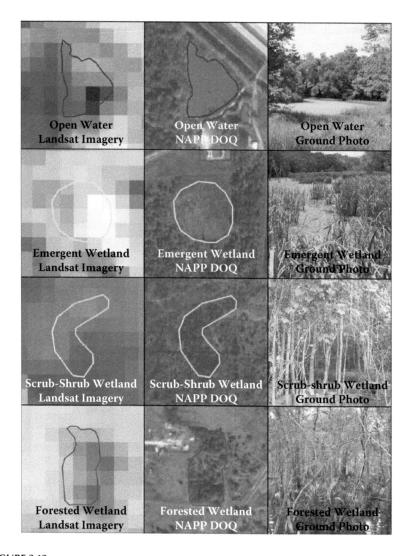

FIGURE 3.12
Palustrine wetland classes were distinguished from each other using Landsat satellite data,
after initial mapping of all wetlands and water in the study area.

(RuleQuest Research, v. 1.8) was used to construct regression tree classifiers
for the data sets. The CART Classifier of the Imagine CART module was
used to implement the regression tree classifier produced by the See5 software package and create a wetland characterization map.

Thematic Classification

Palustrine wetlands are a *system* and defined by Cowardin et al. (1979) as all nontidal wetlands dominated by trees, shrubs, persistent emergents, emergent mosses or lichens, and all such wetlands that occur in tidal areas where salinity due to ocean-derived salts is below 0.5 ‰ ... includes wetlands lacking such vegetation, but with all of the following four characteristics: (1) area less than 8 ha (20 acres); (2) active wave-formed or bedrock shoreline features lacking; (3) water depth in the deepest part of basin less than 2 m at low water; and (4) salinity due to ocean-derived salts less than 0.5 per cent.

The hierarchical classes under the palustrine wetland system (as described by Cowardin et al. 1979) were used to thematically classify the following palustrine wetland types in the study area: (1) emergent palustrine wetlands, (2) scrub-shrub wetlands, and (3) forested wetlands.

The classification procedure involved a random point sampling procedure. Only sample points falling within NWI palustrine wetland polygons were used to drive the classification process. The CART sampling tool was used to select the final sample points, which were subsequently processed using the See5 data mining tool. The regression tree classification method using the boost option in See5 was used for the final thematic classifications. Wetland areas that were adjacent to rivers, streams, lakes, or marine waters were masked from subsequent analyses using the National Hydrography Dataset (USGS 1999).

Landsat ETM+ Band 5 data (sensitive from 1.55 to 1.75 µ) is strongly absorbed by water and was therefore used to discriminate water versus all other land cover. A simple threshold procedure was used to create binary water versus other areas for each seasonal set of data. It was found necessary to stratify the five-county region into coastal versus more upland forested regions, as some fully forested areas in the upland region exhibited a dark waterlike response, possibly the result of different species composition (and appearance) versus forest in the coastal region. All areas not classified as either water or wetland were assumed to be upland and annotated as such.

To accomplish the detailed analytical purposes of the study of palustrine wetlands, riverine (i.e., lotic), lacustrine (i.e., deep lentic), and marine areas of water were removed through visual inspection and automated selection. These areas were selected using an interior starting point and *region growing* procedures, then deleted from the final water map. Subsequent visual editing was also required to remove some cloud and large building shadows (e.g., downtown Houston). Each season was individually mapped and the results added together to create a composite image (scaled 0 to 4 for number of times water was present among the dates in Table 3.6). The resultant areas of water were assumed to be those palustrine wetlands described by Cowardin et al. (1979, p. 103) as *open water* (e.g., rock bottom, unconsolidated bottom, aquatic bed, rocky shore, and unconsolidated shore).

Because of the extremely ephemeral nature of the open water in palustrine wetlands, it may be possible to improve the classification results by evaluating the optimal discrimination level. Specifically, some water was agricultural field irrigation or temporary flooding conditions, and it should be possible to get the water classification accuracy higher than the 69% to 89% range reported later.

Accuracy Assessment Methods

A team comprised of three senior imagery analysts, an image processing specialist, and a GIS specialist conducted the accuracy assessment of the satellite-derived land cover map. The accuracy of wetland maps was assessed using a reference map generated by the team. The reference map was subsequently converted into a raster grid format for accuracy assessment using digital technology. A set of two error matrices and associated statistical results were generated to clearly describe map accuracy.

Generating the Reference Map

An initial five-class wetland reference map was produced through the photointerpretation of NAPP color infrared imagery, using eighteen stereopairs (Figure 3.11). In order to assess the satellite-derived map, the reference map was produced with the same classification scheme as was used for the map. In addition, because the reference map supplies *ground truth* information for assessing the map, every effort was made to produce the most accurate reference map possible. Thus the initial reference mapping was performed by a senior imagery analyst with extensive wetland mapping experience. The analyst, a professional imagery analyst with thirty years of wetlands mapping experience, resided in Texas and has extensive experience mapping wetlands in the Texas coastal area. The independent reference map was produced using standard aerial photographic interpretation techniques (Philipson 1997). A zoom stereoscope system and collateral data, such as soil surveys, were used to interpret the aerial photographs and recorded in analog format. Findings were ink-transferred onto overlays to 1:10,000 scale enlargements of digital orthophotos (Figure 3.13) that were available from the 1995 CIR imagery (i.e., the same imagery that was viewed in stereo). To ensure that the reference map was as accurate as possible, the other two senior image analysts independently reviewed the map and identified errors, which were then corrected through consensus.

Error Matrices and Statistics

Two accuracy assessment error matrices and statistical results were produced to assess the thematic accuracy of the land cover map produced through the processing of Landsat ETM+ data. Because the error matrices were to

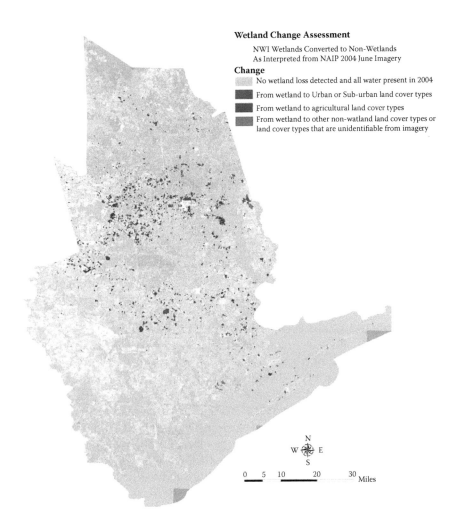

FIGURE 3.13
(See color insert.) A wetland change analysis has been completed and is being used to improve the accuracy of the wetland map in the study area. The wetland change results describe the conversion of wetlands (per NWI maps) from the 1980s through summer 2004 (National Agriculture Imagery Program [NAIP] photography). An example of an area of wetland conversion that is currently "unidentifiable": a building that could be used for urban storage or nearby agriculture but could not be further clarified using the available remote sensing data or other geographic information.

be generated with Imagine software in a digital format the analog reference map overlays were digitized into a digital file format (i.e., GIS "shapefiles"). The shapefiles were subsequently converted into a raster *grid* format for final production of the error matrices. The accuracy assessment of the wetland map was performed using standard random sampling, a common sampling methodology used for thematic data (Congalton 1988), stratified by land cover type for each of two accuracy assessments. Experts indicate that a (liberal) minimum of 50 sample sites should be selected (Hay 1979; Congalton and Green 1999) for each of the five land cover classes, which was applied to the data set. Using Imagine software, a total of 6,650 accuracy sample points were selected, inclusive of the five land cover classes. Accordingly, each of the sample points were located within one of the eighteen stereopair reference map areas.

To generate the error matrices the land cover values of each of the 6,650 sampling points from both the satellite-derived wetland map and reference map were recorded. To account for positional errors in aligning the two maps, the most common value in a 3 × 3 window around a sample point was selected for the class value for that sample point. Using these values, error matrices were produced with the reference data columns representing *ground truth* data (as interpreted from the aerial photography) for each sample point and rows representing the satellite-derived land cover map class for each sample point.

Overall classification accuracy, producer's accuracy, user's accuracy, and kappa (KHAT) statistics were produced for each of the two error matrices (Congalton and Green 2009). The overall classification accuracy is a summary measure of the accuracy of the classified land cover classification data. The producer's and user's accuracies provide accuracy information for each classified land cover type. Producer's accuracy is a measure of the omission errors in the classified land cover data set. It measures, for example, the probability that a pixel identified (on the reference map) as emergent wetlands was correctly identified as such in the classified land cover set data. The user's accuracy is a measure of the commission error in classified land cover data set. For example, it measures the probability that a pixel identified (in the classified land cover data set) as forested wetlands is also identified as such on the reference map. The kappa statistic measures the proportion of correctly identified sample points after the probability of chance agreement has been removed (Congalton 1991).

Characterization of Wetlands in the Landscape

The presence and areal extent of palustrine wetlands in the study area was reported within a FEMA floodplain zone among the five counties of the study area. The Q3 Flood Data are derived from the Flood Insurance Rate Maps published by FEMA (FEMA 1996) to provide a basis for floodplain management, mitigation, and insurance activities for the National Flood Insurance Program. ArcGIS (Environmental Systems Research Institute, v. 9.1) functions were used to process the map outputs from the (satellite data classification) wetland map output as database files and converted to Microsoft Excel tables. The resulting tables summarize the wetland counts and areas for palustrine wetlands in the coastal Texas study areas.

Digital Characterization Map Production

The satellite-data-derived five-class wetland map excludes wetlands that were contiguous with streams and rivers (e.g., riverine or riparian wetlands), lakes (e.g., lacustrine wetlands), and marine waters (e.g., estuaries and Gulf Coast coastal wetlands), but distinguishes between open water (e.g., open water portions of vegetated wetland complexes), emergent wetland, scrub-shrub wetland, and forested wetland (Cowardin et al. 1979). Often open water and vegetated portions of palustrine wetlands are interspersed, and therefore can be considered together or separately for mapping, depending on the focus and goals of the map user.

Accuracy Results

Two sets of error matrices, which include producer's accuracy totals, user's accuracy totals, overall classification accuracy, and kappa (KHAT) statistics were generated for the wetland map data. Palustrine wetland subclasses were combined into a single class *wetland* to investigate the overall accuracy of the palustrine wetland mapping (Table 3.8); subclasses of palustrine wetlands were separately analyzed to investigate how each contributed to the accuracy of the map product (Table 3.9). Overall classification accuracy was calculated by summing the diagonal rows (the correct class predictions) in the error matrix and dividing by the total number of sample points. The overall classification accuracies for the palustrine wetland map (Table 3.8) and the separate wetland subclass map (Table 3.9) were 83% and 80%, respectively.

   The data set used to produce Table 3.8, which has only three land cover types (the three wetlands classes having been collapsed into one class), has a slightly higher accuracy than the land cover data sets used to produce the other two matrices, but the map contains less detailed information about wetland classes. Producer's accuracy statistics, a reference-based accuracy, are a measure of omission errors. User's accuracy statistics, a map-based accuracy, are a measure of commission errors. For a given land cover class, the producer's accuracy and user's accuracy can be quite different. For instance, forested wetlands (Table 3.9) have a producer's accuracy of 53% and user's accuracy of 94%. Thus it is important to view the producer's and user's accuracy together to get a clear picture of the accuracy of the wetland map and to also compare these results to the combined-wetland class accuracies in Table 3.8.

   The kappa statistic (KHAT) for each of the two error matrices is a comparison of how well a land cover data set measures reality versus a randomly generated data set. Table 3.8 (palustrine wetland, combined) has an overall KHAT of 59% and Table 3.9 (wetland subclasses, separated) has a KHAT of 55%, which is a description of how much more accurate the data are than for a randomly assigned land cover value for each image pixel for the

TABLE 3.8

(a) Accuracy Assessment and (b–c) Statistical Results for All Wetland Classes (Combined), Excluding Water Bodies in the National Hydrography Dataset

(a) Reference Data

Classified Data	Upland	Water	Wetlands	Row Total
Upland	4176	10	731	4917
Water	44	158	24	226
Wetlands	338	9	1160	1507
Column Total	4558	177	1915	6650

(b)

	Reference Totals	Classified Totals	Number Correct	Producers Accuracy	Users Accuracy
Class Name	4558	4917	4176	92%	85%
Upland	177	226	158	89%	70%
Water Wetlands	1915	1507	1160	61%	77%
Totals	6650	6650	5494		

Overall Classification
  Accuracy = 83%

(c)

Overall Kappa Statistics =
  0.5929

Conditional Kappa for each
  category:

Category	Kappa
Upland	0.5210
Water	0.6909
Wetlands	0.6766

combined-wetland class (i.e., 4%). Accordingly, to summarize the accuracy of the map results: forested palustrine wetlands are most accurately mapped, followed by all water (i.e., including palustrine open water wetlands), emergent palustrine wetlands, and scrub-shrub palustrine wetlands, in that order (Table 3.9). The map accuracy described for this case study was consistent or exceeds that of other wetland mapping products that use manual and automated mapping techniques, including some wetland map products that use finer resolution data than the Landsat ETM+ data. However, certain wetland classes in the map have accuracies that were less than expected, and additional methods have been applied to improve these accuracy issues (e.g., for emergent wetlands).

TABLE 3.9

(a) Accuracy Assessment and (b–c) Statistical Results for Wetland Classes (Analyzed Separately), Excluding Water Bodies in the National Hydrography Dataset

(a) Reference Data

Classified Data	Upland	Water	Emergent Wetland	Scrub-Shrub Wetland	Forested Wetland	Row Total
Upland	4171	19	135	26	673	5024
Water	71	229	14	8	9	331
Emergent Wetland	196	13	70	4	10	293
Scrub-Shrub Wetland	27	10	32	7	48	124
Forested Wetland	43	2	4	5	824	878
Column Total	4508	273	255	50	1564	6650

(b)

Class Name	Reference Totals	Classified Totals	Number Correct	Producer's Accuracy	User's Accuracy
Upland	4508	5024	4171	93%	83%
Water	273	331	229	84%	69%
Emergent Wetland	255	293	70	27%	24%
Scrub-Shrub Wetland	50	124	7	14%	6%
Forested Wetland	1564	878	824	53%	94%
Totals	6650	6650	5301		

Overall Classification
 Accuracy = 80%

(c)

Overall Kappa statistics
 = 0.5521

Conditional Kappa for
 each category:

Class Name	Kappa
Upland	0.4729
Water	0.6787
Emergent wetland	0.2086
Scrub-shrub wetland	0.0493
Forested wetland	0.9196

Characterization and Interpretations

The accuracy-assessed maps produced in this effort inform wetland functional assessments can be used to compare with wetlands mapped in the past using NWI to, to better understand changes that have led to the present landscape, allowing for an assessment of the drivers of wetland change (Figure 3.13).

TABLE 3.10

Number of Palustrine Wetlands Within and Outside of the FEMA Floodplain Zones

County	Open Water	Emergent Wetland	Scrub-Shrub Wetland	Forested Wetland
(a) Number of Palustrine Wetlands within the 100-Year FEMA Floodplain				
Brazoria	7186	25961	28590	19889
Fort Bend	1726	4888	8149	8719
Galveston	4082	10040	3804	1378
Harris	3719	8071	10719	7756
Montgomery	1769	2909	11294	10324
(b) Number of Palustrine Wetlands within the 100- to 500-Year FEMA Floodplain				
Brazoria	821	8083	6470	4936
Fort Bend	385	902	1920	1954
Galveston	550	3394	1790	676
Harris	1382	2212	2666	1784
Montgomery	654	832	3111	3173
(c) Number of Palustrine Wetlands outside the 500-Year FEMA Floodplain				
Brazoria	2875	25623	19384	16440
Fort Bend	1971	9661	3749	4552
Galveston	786	9171	6538	3340
Harris	5584	9417	8629	4886
Montgomery	3754	6904	14678	13624

One method for estimating the hydrologic functions of the mapped wetlands was to compare the likelihood of flooding within specific areas of the study area to the presence of palustrine wetlands. Because many of the mapped palustrine wetlands have small, ephemeral, few, or an apparent lack of channels that connect them to surface water bodies or other palustrine wetlands, it might be concluded that these wetlands are never connected to other surface water bodies (i.e., are *geographically isolated* from other water bodies). We used FEMA flood zone data to quantify the potential for hydrologic connectivity between palustrine wetlands and other surface water. Specifically, we hypothesized that overland *sheet-flow* during a flood event or connections by way of very narrow channels may be indicated by FEMA flood zone determinations because the base data were of sufficiently fine resolution to discern small channels and to model the maximum extent of potential flooding events (FEMA 1996) in the vicinity of palustrine wetlands, which may have very little topographic relief depending upon the region of the study area.

We estimated periodic connectivity of palustrine wetlands with flooding surface waters by quantifying the presence of palustrine wetlands within 100-year (Table 3.10a) and 100- to 500-year (Table 3.10b) FEMA (1996)

TABLE 3.11

Size of Palustrine Wetlands in FEMA Floodplain Zones

County	Open Water	Emergent Wetland	Scrub-Shrub Wetland	Forested Wetland
(a) Area (Hectares) of Palustrine Wetlands within the 100-Year FEMA Floodplain				
Brazoria	8847	712	2805	2081
Fort Bend	3933	877	1243	8789
Galveston	18106	20170	688	255
Harris	11008	2389	1407	2903
Montgomery	29165	50791	5657	19695
(b) Area (Hectares) of Palustrine Wetlands within the 100- to 500-Year FEMA Floodplain				
Brazoria	345	90	393	380
Fort Bend	238	130	250	1733
Galveston	523	1595	454	91
Harris	372	277	258	171
Montgomery	1027	3419	1477	3438
(c) Area (Hectares) of Palustrine Wetlands outside the 500-Year FEMA Floodplain				
Brazoria	1673	899	2219	1661
Fort Bend	1309	2872	611	2122
Galveston	59482	3545	1837	626
Harris	7613	2206	1116	847
Montgomery	2856	8746	3779	9031

floodplain zones by wetland subclass. Depending on the county and the subclass of palustrine wetland in the study area, palustrine wetlands (by count) were located within the 100-year floodplain zone (Table 3.10a), within the 100- to 500-year floodplain zone (Table 3.10b), or outside of the 100- to 500-year floodplain zone (Table 3.10c). The trends among counties and for some of the wetland subclasses can probably be best understood in the context of the gradients in terrain from the coastal plains of Brazoria County and Galveston County to the hillier regions of Montgomery County, and in the context of historical and contemporaneous land cover or land use.

We estimated the potential (surface-water related) hydrologic functions of palustrine wetlands in the study area, with regard to surface water by quantifying the area of palustrine wetlands within 100-year (Table 3.11a) and 100- to 500-year (Table 3.11b) FEMA floodplain zones by wetland subclass. Depending on the county and the subclass of palustrine wetland in the study area, palustrine wetlands (by area) were located within the 100-year floodplain zone (Table 3.11a), within the 100- to 500-year floodplain zone (Table 3.11b), or outside of the 100- to 500-year floodplain zone (Table 3.11c).

These trends were different for some of the counties and for some of the wetland subclasses, and can be best understood in the context of historical and contemporaneous land cover or land use among the counties of the study area. Any wetland that falls among two or more floodplain zones (e.g., part of the wetland in the 100-year floodplain and another part of the same wetland in the 100- to 500-year floodplain) was included in both floodplain zones by count and by total wetland area. This provides an assessment of wetlands in each of the floodplain zones; however, it overestimates the total number and total wetland area if the individual county values are summed. Therefore, summing the values in this case study for different counties will result in an overestimation of wetland count and area values, and is not recommended. These mapping results must also be understood in the context of the data quality and the level of data processing that has been conducted to date. At this stage of research, several irregularities have been observed for the distribution of wetlands among counties, which are described next. Aside from the traditional accuracy assessment, the map analyses have been compared to the current understanding of wetland distributions among the counties in the study area.

Using Landscape-Scale Wetland Characterization Lessons Learned

The culmination of the approach described in this case study is a characterization map and associated accuracy assessment results. These results are extremely valuable and are useful for placing the characterization in perspective for users and decision makers, who need a guide on using the outputs appropriately, and should be placed in the context of the overall improvement of knowledge about the wetlands of the area. These results also point out key research areas and knowledge gaps that can then be addressed in follow-on work. The results of this case study demonstrate this approach by pointing out that:

1. The total area of palustrine open water wetland in Galveston County was greater than expected. We hypothesize that the greater than expected acreage of open water was a direct result of our other-than-palustrine-wetland extraction technique, which used the National Hydrography Dataset (NHD), which contains geographic information about streams, rivers, and lake boundaries. It is possible that the NHD does not accurately represent all the streams, rivers, and lakes in Galveston County because they were below a size threshold or for other unknown reasons. We also hypothesize that, because Galveston County was partially bounded by the Gulf of Mexico, the Landsat ETM+ band thresholding of coastal waters did not sufficiently exclude tidal wetlands that exceed 0.5‰ salinity, per the Cowardin et al. (1979) definition of palustrine wetland.

2. The total area of palustrine open water wetland in Montgomery County was greater than expected. Similar to Galveston County, we

hypothesize that the greater than expected acreage of open water was a direct result of our other-than-palustrine-wetland extraction technique, which used the National Hydrography Dataset (streams, rivers, and lake boundaries). It is possible that the National Hydrography Dataset does not accurately represent all the streams, rivers, and lakes in Montgomery County because they were below a size threshold or for other unknown reasons.

3. The area of forested palustrine wetland in Fort Bend County was substantially greater than that in Brazoria County. We hypothesize that these results may be the result of our method of mapping forested areas, that is, mapping those areas that were wet during any of the seasons or years of the Landsat data used. The relatively liberal selection approach of including wet forested areas may have resulted in an overestimation of uplands woods in Fort Bend County that were wet at the time of Landsat data acquisition. It may also be that masking of coastal nonpalustrine waters caused an underestimation of the area of very wet palustrine forest in Brazoria County (e.g., large swamps with open canopies, near the Gulf Coast).

4. The area of emergent palustrine wetland in Galveston County was substantially greater than that in Brazoria County. We hypothesize that this may be the result of the predominance of forested areas in Brazoria County relative to Galveston County. It may be that emergent (palustrine) wetlands were differentially confused for forested palustrine wetlands in Brazoria County relative to Galveston County during the automated Landsat classification process.

The probability that a pixel identified (on the reference map) as a palustrine wetland or a specific type of palustrine wetland, along with the kappa statistic, provides the likelihood of actually going to a mapped wetland and finding the mapped class to be correct. Thus the likelihood of arriving at a true wetland is greater for mapped wetlands with greater accuracy percentages and greater kappa statistic values. It is, however, important to consider that the most accurate way to identify wetlands to class on a wetland-by-wetland basis is to manually interpret fine-scale imagery or to physically visit the wetland sites on the ground.

Because the number of palustrine wetlands in the study area is approximately 400,000, the task of either manually interpreting fine-scale imagery or field surveying these sites (or sampling a statistically representative number of sites) is impractical. Thus the results of this case study suggest that practitioners exercise caution when interpreting these results at a fine scale or when drawing conclusions about the ecological condition of specific wetlands based upon these results. However, the use of these results provide a better understanding of the general spatial trends of palustrine wetlands at a broad scale, within the study area, using the interpretation approach discussed in this case study.

The functional assessment of palustrine wetlands requires that there be an accurate basis for the description of landscape position, size, physical dimensions, and temporal changes (e.g., hydroperiodicity) of wetlands. This mapping and accuracy assessment effort furthers the important goals of determining the functional characteristics of palustrine wetlands in the Texas coastal region. Because the concept of *geographic isolation* requires (by definition) the precise description of the physical interrelationships between wetlands and other entities within the landscape (e.g., the distance between a palustrine wetland and surface water bodies), this case study directly focused on the geographic position of palustrine wetlands (and subclasses thereof) relative to surface water and accounts for fluctuating hydrology. Much is still unknown, however, about the comprehensive functional character of these palustrine wetlands. Understanding the location of palustrine wetlands and their ecological subclasses is the first step toward understanding the hydrologic and other more complex ecologic functions of palustrine wetlands, particularly at a broad scale (e.g., the cumulative functions).

Fundamental outcomes of this case study were the (accuracy-assessed) answers to the questions regarding palustrine wetlands in the study area, such as: Where are they? How many are there? What is their acreage? What is their functional role? Although the five counties in the study area are quite different from each other in a variety of ways, we found that each county currently contains a large number of relatively small palustrine wetlands. These relatively small, but numerous, palustrine wetlands cumulatively comprise hundreds of thousands of wetland hectares in the five-county (16,086 km^2) area. Considering that similarly small palustrine wetlands are ubiquitous throughout the total land area of the United States (9.2 million km^2), palustrine wetlands likely comprise a tremendous cumulative area of wetland ecosystem in the United States, with many of the same ecological functions as other more prominent (i.e., typically large single) wetland areas, such as riparian wetlands, lacustrine wetlands, and estuarine wetlands.

The results of this case study demonstrate one way that these relatively small palustrine wetlands can be mapped and assessed for their ecological functions, particularly at a moderately broad scale, which paves the way for an even broader scale ecological assessment of palustrine wetlands in a cost-efficient manner, providing a relatively inexpensive alternative to traditional field surveying, photogrammetric, or semiautomated mapping efforts.

4

Integrating Field-Based Data and Geospatial Data

Effectively Utilizing Data Variables

Decisions about which type of ecological information, remote sensing data, and GIS data to use to begin a wetland characterization (or assessment) landscape indicator development project are difficult because they require an optimization of three important factors: (1) the cost of data (i.e., acquisition, processing, and storage); (2) the availability of the necessary data; and (3) the quality of the data. In the planning of a landscape ecological assessment, whether for ecological research or for regulatory support, one has to decide, for example, between the use of an objective data source with high quality but many gaps in coverage, which would require a large portion of available resources to collect sufficient data, and the use of other data sources with fewer gaps and less costly information, but requiring a reduction in reliability and comparability.

Data Availability, Cost, and Quality

Decisions about how to evaluate and monitor the ecological condition of Great Lakes wetlands must take into consideration the logistical challenges presented by large landscapes and the fact that assessment and monitoring schemes must be parsimonious. Although there are many benefits associated with exploiting existing data, there are costs (e.g., noncontemporaneous data incompatibility) that must be considered in accessing those data. Long-term or wide-area data are generally accessible through major data centers. Short-term or single-site data sets, generated to address focused scientific questions, are often available solely from the originating organization. The overall cost of any landscape assessment is affected by the availability of existing data, its source (whether from a public, nonprofit agency, or from private for-profit companies), and its quality.

Existing data are often fragmented and dispersed among many sources, depending on the geographic and environmental areas that are considered. This issue can be especially relevant when local information is needed at a broad scale for land management, such as the case of coastal wetlands of the entire Great Lakes Basin (Chapter 5). In addition, databases have been created in several formats or geographic projections that may not be interoperable. These are important issues to understand prior to selection of the data or metric.

U.S. federal agencies are likely to be the primary lower cost sources for data that include maps of elevation, watershed boundaries, road and river locations, human population, soils, land cover, and air pollution. Sources include the U.S. Army Corps of Engineers; National Oceanographic and Atmospheric Administration; U.S. Geological Survey; the U.S. Environmental Protection Agency; the U.S. Department of Agriculture; the U.S. Census Bureau; and the Multi-Resolution Land Characteristics Consortium. Resources may consist of databases, raw or preprocessed remote sensing data, digitized maps, and GIS/statistical models or software. Data types that use standard methodologies, such as those that comply with the Federal Geographic Data Committee (FGDC), provide data for which reliability can be high, availability is assured, sources are identified, and good comparability exists. In the absence of these data quality assurances, large quantities of data may be available at lower costs, but with an increased risk of noncomparability, low reliability, and uncertain future availability.

Key Data Types

The types of data that may be available for landscape indicator development include ecological information, remote sensing data, and other geospatial information. Ecologists have traditionally used historical maps, aerial photographs, and their understanding of spatial relationships between ecosystem patches to explore relatively broad-scale ecological characteristics of the landscape (e.g., Miller and Egler 1950; MacArthur and Wilson 1967; Howard 1970). Airborne digital data are useful for determining the abiotic conditions of Great Lakes coastal wetlands (e.g., Lyon and Drobney 1984; Lyon and Greene 1992; Lyon et al. 1995). The inherent complexity of wetland ecosystems and the particular ecological processes of coastal wetlands have also prompted development of, and research into, the use of specialized airborne and satellite sensors, and related processing techniques for these new data types. Within the past decade, environmental scientists have successfully integrated and applied the use of relatively sophisticated sensors (e.g., airborne hyperspectral, airborne multispectral and satellite multispectral scanners), automated image processing software and techniques (e.g., Imagine software, http://geospatial.intergraph.com/products/ERDASIMAGINE/ER DASIMAGINE/Details.aspx; ENVI software, http://www.exelisvis.com/ ProductsServices/ENVI.aspx), and the computing power of GIS (e.g., ESRI's ArcView and ArcInfo, www.esri.com; accessed July 21, 2012) to the study of ecology. However, most of these tools are relatively new, and the application of multiple sensors and techniques to wetland detection and analysis has not fully matured.

Ecological Information

It can be possible to sample ecosystem components, such as the water, soil, air, plants, and animals, but it can be impossible to survey (i.e., fully measure all of these components throughout the entire ecosystem). Thus any ecological study must consider the components of an ecosystem to answer the question of which ecological indicators are the best surrogates of overall ecosystem condition. A traditional ecologist might consider a good coastal wetland indicator as one that can be explained in its component parts, which is known a priori to be directly linked to the functional status of wetlands (e.g., hydrology), and has a demonstrable and repeatable linkage with the functional status of the wetland. A landscape ecologist might additionally consider a good wetland indicator as one that reflects conditions across multiple wetlands, multiple watersheds, multiple lake basins, or at other more broad scales. Environmental policy experts may be additionally interested in selecting ecological indicators that have the greatest likelihood of answering the following related questions:

1. What indicators characterize and measure ecological sustainability?
2. What indicators best show changes caused by human impacts?
3. How can indicators developed in one place and time be used in other places and times?

To some extent, different measures and monitoring designs are needed to answer all of these questions, and to answer them at local, watershed, regional, national, and basinwide scales. Although local or watershed assessments may include fairly complete monitoring of stresses and impacts, such direct assessment is not practical over large regions. But there are opportunities to harmonize assessments across spatial scales by including, together with field monitoring, advanced and less expensive assessment methods that utilize remote sensing data.

Associating Ecological Data and Remote Sensing Data

Because of the vast areas involved and the complexity of information that is required to assess the ecological functions of large areas like the Great

Lakes Basin, remote sensing technologies have been developed to provide an additional source of information to develop indicators of coastal wetland condition. However, our ability to interpret landscape spatial patterns and identify the materials on the ground can be a challenge because of the limits of the spectral (e.g., detectable energy wavelengths) and spatial (e.g., minimum pixel size) properties of the sensor. Remote sensing data are often mistakenly thought to be less useful than observations that are made on the ground within ecosystems of interest. However, field-based measurements (e.g., 1,000 sample points within a wetland) may not be as comprehensive as remote sensing data and thus may be relatively less effective at determining the true type, number, and distribution of some of the key elements of an ecosystem (e.g., wetland plant species distribution). Remote sensing data can supplement the inability of investigators to effectively sample a wetland by providing *wall-to-wall coverage* (e.g., a complete coverage; a full image that assesses wetlands from edge to edge of the ecosystem).

The inherent trade-off of having a full coverage of remote sensing data, rather than a field sample, is being distant from the wetland and consequently having the (satellite or airborne) image picture elements (pixels) at sizes that are coarser grain than field-based observations, but having a complete coverage of these data. This is the reason why remote sensing data (or derived geospatial data) should be thought of as a supplement to field-based information and not as a replacement for field-based information.

It is important to determine the scale of ecological information that is necessary to assess the ecological condition of coastal wetlands prior to determining which types of remote sensing data are required for the assessment. For example, if the goal is to measure simple spectral characteristics (e.g., measuring for the presence and area of large open water areas), then it is not necessary to have fine-scale spectral or spatial resolution information, and relatively coarse resolution satellite data (e.g., MODIS or Landsat) should suffice. If the goal is to measure more complex characteristics of coastal wetlands (e.g., the presence and location of all emergent vegetation, combined), then it may be necessary to acquire data with higher spatial resolution for those areas of vegetation (e.g., a 50 m wide patch of vegetation on the edge of a marsh) that can still be used to cost-effectively cover a vast area. The appropriate spatial resolution of remote sensing data is thus determined by the investigator deciding what the minimum conceivable level of spatial information is necessary to adequately assess coastal wetlands and then determining the optimal pixel size that provides that information, in the context of the cost and availability factors. The spectral resolution necessary for the landscape assessment is determined similarly by the investigator deciding what the minimum conceivable level of spectral information (e.g., how much information must be directly extracted from the reflectance of plant leaves) is necessary to adequately assess the condition of wetlands and to address the ecological endpoints. The ecological investigator should also determine whether there is a need for temporal data analyses in the future and at what frequency the remote sensing acquisition might be required to adequately assess the condition of coastal wetlands. All remote sensing platforms (e.g., airborne or satellite equipment) have differing return rates (i.e., repeated overflight), ranging from approximately daily to once every several weeks.

Because the characteristics of land cover or land use in the vicinity of wetlands may determine the condition of the wetland, and these conditions may change over time, it is often desirable to assess these features repeatedly over time using remote sensing. Stresses related to land cover and land use may be those directly caused by humans, such as agricultural, urban, and industrial development. Other human-induced disturbances in coastal wetlands include those associated with upland development, shoreline development, deforestation, changes in upland agricultural practices, road construction, dam construction, or other hydrologic alterations. These changes or activities can be directly observed or inferred using remote sensing. Aerial photographs (generally 1 m spatial resolution) or airborne digital imagery (generally a 1 to 5 m spatial resolution) are often available for specific areas and can be used to correlate land-use changes with wetland alterations.

Land use and land cover data are most often derived from some type of overhead remotely sensed imagery, such as aerial photographs, airborne digital data, or satellite digital data. Data collected by satellites are most often used to map land cover over vast areas and have been used to measure changes over time. With a few exceptions, most of the sensors carried on satellites measure light reflected from the earth's surface. Because different surfaces reflect different amounts of light at various wavelengths, it is possible to identify general vegetational change (Figure 4.1) or broad land cover types (Figure 4.2) from a variety of airborne and satellite measurements of reflected light among multiple scales.

Remote sensing data are the source for much of the derived land cover and land use data sets that are used for GIS analyses and modeling. Generally, GIS products are derived by either manual remote sensing data interpretation or semiautomated image processing. The example in this chapter includes indicator development using a variety of spatial and spectral resolution digital remote sensing data relying substantively on field-based data.

Geospatial Data

Geographic information describes the locations of landscape entities and can be interpreted so that the spatial relationships between these entities are understood. Most of the broad-scale geographic information produced today resides with national and state governmental groups but is frequently produced at fine-to-moderate scales by local governments, individuals, corporations, and other nongovernmental organizations. All of the geographic information products from the United States that are used in this chapter can be readily downloaded to your personal computer by visiting the website URLs that are listed in this chapter. A computer and GIS software are

FIGURE 4.1
Airborne multispectral imagery (e.g., Positive System's 1 m spatial resolution, 4-band ADAR data) can be utilized for specific areas and may be used to correlate land use or land cover changes with wetland alterations, at local to regional scales, such as is shown here in the Wildfowl Bay region of Saginaw Bay, Michigan (Lake Huron).

FIGURE 4.2
(See color insert.) Different surfaces reflect different amounts of light at various wavelengths so it is possible to identify general vegetational change from satellite measurements of reflected light as depicted here for the entirety of the U.S. and Canada Great Lakes Basin. (Courtesy of Burt Guindon, Canada Centre for Remote Sensing.)

required to process and analyze these digital geographic data sets (e.g., using Geographic Resources Analysis Support System, http://grass.fbk.eu/mirrors.php; or Environmental Science Research Institute's ArcInfo or ArcView software, www.esri.com; accessed July 21, 2012).

The categories of geographic information cover a wide range of parameters that have been mapped (in what is sometimes referred to as a *thematic map*) at a variety of scales and may include information about topography, human population, land cover, land use, oceans, rivers, streams, lakes, wetlands, roads, important political boundaries (e.g., counties or townships), and features of importance (e.g., national parks, monuments, and landmarks). The level of detail described for each parameter within a map (e.g., type of wetland vegetation—herbaceous and woody plants, or by plant species) is dependent upon the spatial and spectral characteristics of the remote sensing data from which the geographic information or geospatial model was derived.

In a typical thematic map, data are digitally stored as a series of numbers that produce a map of these values associated with each pixel in the map. These maps can be thought of as checkerboards, where each grid pixel represents a data value for a particular landscape characteristic or *theme* (e.g., a map's topographic theme with a point elevation value and pixel value of "245," which defines that particular pixel at 245 feet above sea level). A GIS can be used to view and measure landscape metrics or indicators, using a variety of methods. One method called *overlaying* simply examines several different themes to extract information about the spatial relationships among the themes. For example, by overlaying maps of land cover and topography, the analyst can look at the occurrence of wetlands across the landscape, using an overlay of land cover (which includes wetland locations) on topography (which includes elevational change, i.e., slope, across the entire landscape). These relationships can be digitally stored as a new map, which combines the information from the original set of thematic maps. Another method called *spatial filtering* can be thought of as using a *window* to calculate values within small areas that are part of a larger map. Spatial filtering can be used to create surface maps of metric or indicator values that help to visualize the spatial patterns of metrics or indicators in more detail than is provided, for example, by watershed scale summaries.

Because landscape ecological research involves the use of several GIS data sets, a thorough understanding of how these GIS data sets can be integrated and managed is important during the early stages of the research. With the rapid growth in GIS software and applications, the environmental scientist's capability for storing, manipulating, and visualizing geographic information is becoming commonplace for understanding ecological data, which has shifted some of the emphasis of landscape ecology from the GIS applications and manipulations (described in the prior paragraph) to the statistical analysis of the geospatial data. Such improved geospatial data analyses of the relationships between landscape condition and the ecological functions of wetlands can be enhanced by targeted (i.e., nonextensive) site-specific assessments, allowing for a broader-scale spatial analysis, given a well-designed field, remote sensing, and GIS mapping approach.

Statistical procedures can improve the understanding of the broad-scale relationships between landscape condition and the condition of wetlands that reside there, thereby allowing for larger data sets to be analyzed using analytical techniques that allow for the inclusion of data that might otherwise cause analysis difficulties. Such geospatial statistical techniques have demonstrated initial significant relationships between wetland parameters and other mapped geographic data in the vicinity of these wetlands, but the strengths of the relationships can be variable, and the causal relationships are somewhat uncertain at this time.

The potential limitations of using mapped geographic data to assess wetlands are directly related to the capability of linking GIS-based assessments to relevant field-based assessments of wetlands (Whigham et al. 2003), an important component of determining the accuracy of landscape indicators of wetlands.

Metric Measurability, Applicability, and Sensitivity

After reviewing data availability, the next step in a landscape assessment is the selection of landscape metrics, which requires considering three important questions:

1. Are the available data capable of adequately measuring the (metric) parameters, and do they address the ecological endpoints of interest?
2. Are the metrics to be derived during the landscape ecological analyses applicable to the ecological endpoints of interest, and do these results answer the questions of the audience for the analyses?
3. Are the metrics to be derived during the course of the landscape ecological analyses likely to be sensitive enough to provide information about the ecological endpoint(s) of interest?

Evaluation of measurability of a landscape metric must include a primary review of the expertise, training, and methodologies used to acquire and process the remote sensing data, input and analyze the derived GIS data, and synthesize the results of such analyses. These four measurability-related steps require the input from individuals that have expertise in remote sensing, computer science, geography, GIS, wetland ecology, general ecology, environmental science, chemistry, hydrology, geology, and other relevant specialized fields. Evaluating measurability also involves an early review of prior techniques, and determining how they may be modified to accomplish the particular goals of a landscape scale wetland assessment. It is often tempting to repeat some of the same techniques applied in other geographic locations or in other ecosystem types, but many of the techniques in GIS and remote sensing work do not apply to wetland assessments. One of the principal differences between landscape-scale wetland assessments and assessments of other ecosystem types is that wetlands are, by definition, a transitional zone between upland and open water ecosystems. Thus there is a mixture of upland and open water conditions at different times in wetlands, which may lead to interpretation confusion if solely upland or solely open water methodologies are applied to wetland ecosystems.

The applicability of a landscape metric (or indicator, derived from that metric) is also a critical step that a research team should address prior to beginning the landscape assessment process. For example, in the Great Lakes region, the State of the Lakes Ecosystem Conference (SOLEC) has compiled several lists of operational and proposed field-based and other measurements that are applicable to landscape assessments that can be used to develop landscape indicators. Not all of these measurements have been completed at a broad scale and some have been completed solely at selected sites around the Great Lakes. Additional information about these measurements may be used as analogues for applicability of landscape scale metrics for the development of landscape indicators in other areas.

Landscape metrics and metric-derived indicators must demonstrate that they are sensitive to (spatial and temporal) changes that occur in wetlands if they are to provide information to the relevant audiences, determined in the *measurability* evaluation. Because all ecological systems change, metric and indicator sensitivity must be gauged at a relevant spatial and temporal scale that makes sense for the ecological endpoint, determined in the *applicability* evaluation. Metric sensitivity can be evaluated by field verification and then further evaluated (as a landscape indicator) by validating a response by the metric to known (and validated) drivers within or in the vicinity of wetland ecosystems. For example, a landscape metric that approximates nutrient conditions in wetlands of two different watersheds can be field tested using a statistically valid field sample of wetland soil and water chemistry in those two watersheds, and then could further be tested as an indicator by field validating the relative proximity of agricultural land to those wetlands throughout each of the watersheds.

The field-based validation of the landscape metrics or indicators is a vital (often neglected) step that requires a statistically sound methodology prior to entering the field to conduct sampling. A typical way of determining if a landscape metric or indicator is sensitive is to hold it to a predetermined standard of acceptability (e.g., for a linear regression or ANOVA, a significance level of $\alpha = 0.05$), but this standard is not always the same for every metric, geography, ecosystem, and relevant audience. This step relies on the a priori knowledge and expertise of the research team and is a standard method for designing a hypothesis test. Thus a critical analysis step is for the research team to agree on the level of sensitivity that is required to satisfy the various research or policy goals of the project prior to the commencement of field validation. The sensitivity of a landscape metric can also be assessed by comparing the results and distribution of the metric with the results of other studies, but such an approach is insufficient for fully developing a landscape indicator and should be avoided if possible.

Categorization

Edited training sets (see Training Data Set Development section below) should be used for categorization of original remote sensing data scenes. The maximum likelihood or Bayesian decision rule approach to categorization has been evaluated and utilized among a variety of applications and geographies. There are a number of reasons to use this approach, among which is the fact that it has been used successfully in remote sensing applications for over twenty years. This approach makes use of both Euclidean distance measurement and probability-based criteria in the determination of the class identity of a given pixel. The use of these criteria makes the categorizing algorithm more sensitive to class characteristics as compared to single-criterion approaches.

After evaluation of training set quality, the image scenes can be categorized using four or more bands as input to the maximum likelihood classification algorithm. A sensitivity experiment or feature selection (Jensen 2004) can evaluate the appropriate bands to be employed. Studies have shown that one should employ only those bands important to identify the land cover types of interest. One can select a subscene of the data of small portion, for example, 250 pixels × 250 pixels, of the whole scene. From the literature, a set of several bands can be selected and this small subset categorized. Comparison with aerial photos or other images allows determination as to the value of the bands that were selected. Iterations of different combinations of bands through the categorization algorithms allow the selection of the optimal combination to identify the land cover types of interest (Lyon et al. 1992).

The resulting product of one hundred or more classes will represent a preliminary land cover (PLC) product. It remains to identify the land cover type of each cluster or class. This activity is known as *labeling* and can be performed by using local knowledge, ground information, aerial photographs or images, and maps. The PLC product requires that certain clusters or classes be aggregated to render the best product. The aggregation will allow inventory of several clusters or classes that may represent the same functional land cover class as represented in the selected land cover categorization system.

Training Data Set Development

The development of training data sets for categorization requires a method that will yield high quality and unbiased categories of landscape characteristics in an automated fashion. The unsupervised training set development approach has been proven successful in providing good, homogeneous spectral clusters in applications such as land cover or landscape characterization. The unsupervised approach negates many systematic problems associated with atmospheric and terrain radiometric distortions. The unsupervised approach has been used successfully for over twenty-five years, and it can also be implemented with a minimum of effort. Consistent results can be expected for most applications.

Unsupervised training set development is conducted with an automated spectral clustering algorithm. A common clustering method can be selected from the available choices in the literature of remote sensing applications. Suitable examples have often been implemented on applications software and hence are available to most users.

Training set development can be conducted with some assumptions and input parameters. In change detection projects, individual images under study should be subjected to unsupervised training set selection or clustering separately. Each image scene can be clustered on an iterative basis for as many as one hundred or more clusters or classes. Criteria for linking or breaking (often referred to as *busting*) of clusters are usually established empirically, and the defaults can be used initially until experience is developed as to the most appropriate conditions for the given application.

The results of unsupervised training set development can be evaluated in several ways. A preliminary categorization image will be produced from the unsupervised training set development or clustering activity. This image will help indicate the location of many clusters or classes. To both identify the type of land cover associated with each class and to evaluate quality, the preliminary categorized image can be viewed individually class by class. Each class can be colored from a table of colors to facilitate analysis.

The preliminary categorized image can be used along with other procedures to evaluate the quality of individual training sets or clusters using aerial images, fine resolution satellite images, or other data. Procedures include divergence calculations, bivariate scatter plots of cluster means and variance–covariance, and comparison of the preliminary categorized image with ground information or aerial photographs.

Following the analyses of the quality of training sets, the contents may be edited to remove examples of poor clusters, for example, clusters that represent very few pixels in the original image, overlapping clusters, and clusters that are composed of more than one distinct, spectrally homogeneous land cover class.

Change Detection Using Principal Component Analysis

Principal component analysis (PCA) involves the reorientation of axes of an input data set creating output principal component (PC) data sets. In the case of change detection, two coregistered images would be input as, say, an eight band or eight axis image. Because of the autocorrelation of the original data, there is an elongated cloud of distribution of data located in the axis of each data set (Lunetta and Elvidge 1998). The PCA will orient the first axis of the output data through the central core of the input data cloud such that the interband variability is maximized. The second axis will be perpendicular to the first and will be directed through the next major direction of variance in the data set such that the intraband variability is maximized for the second axis. The creation of new axes continues until a default number (eight in this case) or limit is reached.

The first principal component explains the major variability in the image and contains the overall scene brightness variations that are in common between all the input bands. The second, third, and fourth (and sometimes higher) PC images frequently contain information on pixels that changed in reflectance between the two dates of imagery. The last PC image would be expected to contain random noise that existed in one image relative to the other.

In scenes with cloud cover and associated shadowing, it will be necessary to screen out the cloud and shadow areas to exclude them from the PCA. Clouds or shadows that are present on one date and absent from the second date will tend to redirect the axes that would otherwise be established due to reflectance changes due to land cover change.

Part of the PCA algorithm involves the normalization or equalization of the input data, thereby reducing atmospheric and sensor radiometric response differences from the output PC images. This is one of the characteristics required for many change detection, making PCA a useful approach in certain analyses.

For interpretation of results, one can make brightness or greenness images using the first three principal components. Developed over several years, the brightness or greenness images facilitate interpretations of cause and effect, and can help identify the location and the spectral characteristics of changed land cover types.

Assessment of Accuracy

The assessment of accuracy of the categorized or classified product is a necessary step to ensure a quality product for subsequent analysis, and to meet quality assurance and quality control (QA/QC) objectives. This can be accomplished by fieldwork or image interpretation of the land cover classifications, and conducted according to the scope or guidelines by a team of image interpreters or field personnel. The image interpretation or field team should utilize an accepted land cover classification system as specified in a scope or a call for proposals document to verify the results of categorization.

Image interpreters should utilize a number of other sources of data, which many studies utilize. The data sources can include color infrared high-altitude photographs; Landsat; SPOT; fine resolution satellite data such as GeoEye image data; aerial and map atlases as available; and U.S. Geological Survey (USGS) quadrangle maps and digital files.

In addition to these sources, the team should utilize any other sources available to the researchers. These include U.S. Department of Agriculture National Resource Conservation Service soil surveys; U.S. Fish and Wildlife Service National Wetland Inventory maps and data; available USGS orthophoto quadrangle products; maps of federal property boundaries; or other source data.

Local groups or companies often have a number of complete and partial county coverages for the area to be studied. These archival photographs can be used along with other data sources to map and classify land cover according to a given scope. These coverages are often at medium and low altitude, and they supply a lot of detail that can support interpretations using high-altitude photographs or satellite products as the base data source. These photographs or images represent a very valuable resource that local groups bring to the project, and the use of these detailed and contemporary photos will both speed the classification work and enhance the accuracy of the type mapping.

Satellite data such as Landsat, SPOT, and GeoEye Image data (for example) can also be processed using image processing systems to produce additional images to assist in interpretation and accuracy assessment efforts. Image-enhanced products can be used to assist photo interpretations, and to help visualize local and regional land cover.

Computer categorizations can be made in a few local areas where data of a given resolution can be of value. It is anticipated that ancillary satellite image data can be helpful in classification of land cover (Figure 4.3) for such themes as wetlands and agricultural types (Figure 4.4). These quality determinations should also be made after the final digital databases or land cover products are produced. In this manner the final polygons and classifications in any project can be checked against the original photographs, supporting data sources, and the original interpreted overlay. With available data, both field-based and synoptic, the practitioner may proceed with the focal work of wetland landscape characterization.

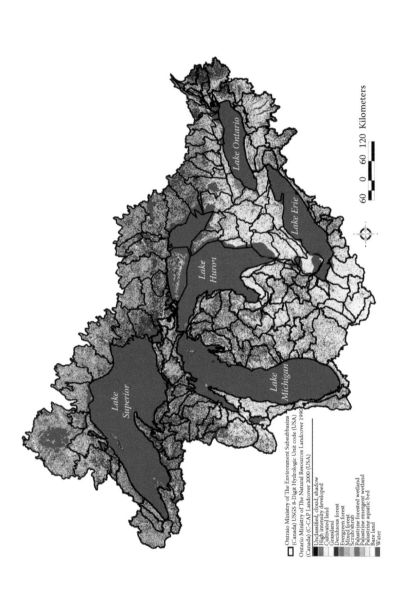

FIGURE 4.3

(See color insert.) Land cover in the Great Lakes Basin as seen from space, using a detailed classification scheme that is a combination of the U.S. National Oceanographic and Atmospheric Administration's 2000s C-CAP land cover and Canada's Ontario Ministry of Natural Resource's 1990s land cover data sets.

FIGURE 4.4
In agricultural areas, lands that do not readily support farming are often used for homes, farm buildings, and cemeteries. Forested areas and wetlands are often adjacent to these domestic areas, as are stream drainages. Here we see a farm area with little natural land cover and a town hugging the drainage way as a place where agriculture is limited.

Case Study: Mapping Invasive Plant Species in the Great Lakes

The aquatic plant communities within coastal wetlands of the Laurentian Great Lakes (LGL) are among the most biologically diverse and productive ecosystems of the world (Mitsch and Gosselink 1993). Coastal wetlands are also among the most fragmented and disturbed ecosystems of the world, with a history of impact from landscape conversions (Dahl 1990; Dahl and Johnson 1991; Dahl 2006). Many LGL coastal wetlands have undergone a steady decline in biological diversity during the 1900s, most notably within wetland plant communities (Herdendorf et al. 1986; Herdendorf 1987; Stuckey 1989). Losses in the biological diversity within coastal wetland plant communities may coincide with an increase in the presence and dominance of invasive (i.e., nonnative and opportunistic) and aggressive (i.e., native and opportunistic) plant species, and parallel those observed in many other ecosystems (Bazzaz 1986; Noble 1989). Research also suggests that the establishment and expansion of invasive and aggressive plant species may be the result of general ecosystem stress (Elton 1958; Odum 1985).

Reduced biological diversity in the plant communities of LGL coastal wetlands may be related to disturbances such as the conversion of land cover (LC) within a wetland or on the edges of wetlands (Miller and Egler 1950; Niering and Warren 1980). These disturbances may include fragmentation by roads, urban development, agriculture, or alterations in wetland hydrology (Lopez et al. 2002). However, little is known about the ecological relationships between disturbance and plant community composition in the LGL, especially at the lake-basin and multibasin scales, perhaps as a result of economic limitations associated with such assessments.

Accurate wetland condition characterization is key to determining larger landscape-scale ecological relationships between (1) the presence and distribution of opportunistic plant species and (2) local landscape disturbance. Remote sensor technologies offer unique capabilities to measure the presence and extent of plant communities over large geographic regions. Thus the goal of the case study was to use ground-based wetland sampling to calibrate airborne hyperspectral data to develop spectral signatures of native opportunistic plant species. LGL plant species of interest included the native common reed [*Phragmites australis* (Cav.) Trin. ex Steud] and cattail (*Typha* spp.), and the opportunistic nonnative purple loosestrife (*Lythrum salicaria*).

The thirteen general study sites included coastal wetlands of Lake Erie, Lake St. Clair, Lake Huron, and Lake Michigan (Figure 4.5). This chapter presents study results and techniques used at one of the thirteen coastal wetland sites, Pointe Mouillee (Figure 4.5, inset), on two of the plant species of interest, *Phragmites australis* and *Typha* spp. Analyses were specifically conducted to assess the utility of airborne multispectral data and airborne hyperspectral data at the remaining twelve sites for detecting relatively homogeneous patches of *Phragmites australis*.

Typically, *Phragmites* communities form large monospecific *stands* that may predominate in wetland plant communities, supplanting other plant taxa (Marks et al. 1994). Compared to other more heterogeneous plant communities, *Phragmites* stands are less suitable as animal habitat and reduce the overall biological diversity of wetlands. From a LGL resource perspective, *Phragmites* is difficult to manage because it is persistent, produces a large amount of biomass, propagates easily, and is very difficult to control with mechanical or chemical techniques. A combined field- and remote-sensing-based approach was used to develop a semiautomated detection and mapping technique to support *Phragmites* monitoring and assessment efforts. Relevant ecological field data provided an important measurable link between airborne sensor data and information about the physical structure of *Phragmites* stands, soil type, soil moisture content, and the presence and extent of associated plant taxa.

Methods

Thirteen coastal wetland study sites were selected from a group of approximately sixty-five potential sites along the coastal margins of western Lake Erie, Lake St. Clair, Lake Huron, and Lake Michigan. Sites were selected using aerial photographs, topographical maps (1:24,000 scale), wetland inventory maps, National Land Cover Data (NLCD), input from local wetland experts,

FIGURE 4.5
Thirteen wetland study sites in the Ohio and Michigan coastal zone, lettered A to M. Sites were sampled during July–August 2001. (Inset image) Magnified view of Pointe Mouillee wetland complex (Site E). White arrows indicate general location of two field-sampled *Phragmites australis* stands. Field-sampled site location legend: Pa = *Phragmites australis*; Ts = *Typha* spp.; Ls = *Lythrum salicaria*; Nt = nontarget plant species; Gc = ground control point. Inset image is a grayscale reproduction of false color infrared IKONOS™ data (Space Imaging, Inc., Thornton, Colorado), acquired August 2001.

and published accounts of coastal wetland studies in the areas (Lyon 1979; Herdendorf et al. 1986; Herdendorf 1987; Stuckey 1989; Lyon and Greene 1992). Site selection criteria mandated that sites (1) generally spanned the gradient of current landscape conditions along the coastline of the lakes, (2) were emergent wetlands (Cowardin et al. 1979), and (3) included both wetlands that are open to lake processes and wetlands protected from lake processes (e.g., diked wetlands or drowned river mouths) (Keough et al. 1999). Sites were selected so that proportions of adjacent land cover generally varied among landscapes in the vicinity of the thirteen sites. NLCD and aerial photographs indicated that site land cover adjacent to all of the study sites included active agriculture, old-field agriculture, urban areas, and forest, in varying amounts. Each of the thirteen selected wetland sites was known a priori to contain at least one of the targeted taxa of interest.

Remote Sensing Data Selection, Acquisition, and Processing

Pointe Mouillee was selected for investigation of the spatiospectral characteristics using both airborne multispectral Airborne Data Acquisition and Registration (ADAR™) System 5500, and hyperspectral PROBE-1™. The ADAR was a four-camera, multispectral airborne sensor that acquires digital images in three visible and a single near-infrared band. Data acquisition occurred on August 14, 2001, at an approximate altitude of 1,900 m above ground level, flying north to south at 150 knots. Average pixel size for ADAR was 75 × 75 cm. Using ENVI image processing software a single ADAR scene in the vicinity of the initial *Phragmites* sampling location was georeferenced (RMS error < 0.06) using digital orthorectified aerial photographs (or digital orthophoto quadrangles [DOQs]) and ground control points from field surveys.

The PROBE-1 is a hyperspectral (whisk-broom) scanner system with a rotating axe-head scan mirror that sequentially generates cross-track scan lines on both sides of nadir to form a raster image cube. Incident radiation was dispersed onto four 32-channel detector arrays. The PROBE-1 data were calibrated to reflectance by means of a National Institute of Standards laboratory radiometric calibration procedure, providing 128 channels of reflectance data from the visible through the short wave infrared wavelengths (440–2490 nm). The instrument carries an onboard lamp for recording in-flight radiometric stability, collected along with shutter-closed (dark current) measurements on alternate scan lines. Geometric integrity of recorded images was improved by mounting the PROBE-1 on a three-axis, gyrostabilized mount, thus minimizing the effects in the imagery of changes in aircraft pitch, roll, and yaw resulting from flight instability, turbulence, and aircraft vibration. Aircraft position was assigned using nondifferential GPS data and tagging each scan line with a time that is cross-referenced to the time interrupts from the GPS receiver. An inertial measurement unit added the instrument attitude data required for spatial geocorrection.

The PROBE-1 sensor had a 57° instantaneous field of view (IFV) that offered better mapping capabilities of vertical and subvertical surfaces than available satellite sensors. The IFV of 2.5 mrad (along track) and 2.0 mrad (across track) results provide an optimal ground IFV of 5.0 to 10.0 m, depending upon altitude and ground speed. Data were collected on August 29, 2001, at an altitude of 2,170 m above ground level at 147 knots (visibility 50 km), resulting in an average pixel size of 5.0 × 5.0 m. Data collection rate was 14 scan lines per second (i.e., pixel dwell time of 0.14 ms) and the 6.1 km flight-line resulted in total ground coverage of 13 km^2. A single PROBE-1 scene at

Pointe Mouillee was georeferenced (RMS error <0.6) with onboard GPS data, DOQs, and GPS ground control points from field surveys using ENVI image processing software.

The single scene of PROBE-1 data was visually examined for missing or noisy bands. After the missing and noisy bands were removed, the resulting 104 bands of data were subjected to a minimum noise fraction (MNF) transformation to determine the inherent dimensionality of image data, segregate noise in the data, and to reduce the computational requirements for subsequent processing (Boardman and Kruse 1994). The MNF transformations, as modified from Green et al. (1988), are cascaded principal components transformations.

The first transformation, based on an estimated noise covariance matrix, decorrelates and rescales the noise in the data. This first step resulted in transformed data in which the noise had unit variance and no band-to-band correlations. The second step was a standard principal components transformation of the *noise-whitened* data. Then the inherent dimensionality of the data was determined by examining the final eigenvalues and the associated images from the MNF transformations. The data space was then divided into two parts: (1) one associated with large eigenvalues and coherent eigenimages and (2) a complementary part with near-unity eigenvalues and noise-dominated images. By solely using the coherent portions, the noise is separated from the data, thus improving spectral processing results (Research Systems, Inc. 2001).

A supervised classification of the PROBE-1 scene was performed using the ENVI Spectral Angle Mapper (SAM) algorithm, a semiautomated processing technique for comparing image spectra to a spectral library. Because the PROBE-1 flights occurred three weeks after field sampling, there was a possibility that trampling from the field crew could have altered the physical structure (thus the reflectance characteristics) of *Phragmites*. For this reason and due to inherent georeferencing inaccuracies of the data within the two *Phragmites* stands, PROBE-1 spectra were collected from a 9-pixel (i.e., 3 pixels × 3 pixels) area centered on the most homogeneous field-verified area within each vegetation stand. The SAM algorithm was then used to determine the similarity between the spectra of homogeneous *Phragmites* and every other pixel in the scene by calculating the spectral angle between them (spectral angle threshold = 0.07 rad). SAM treats the spectra as vectors in an n-dimensional space equal to the number of bands (i.e., a 104-dimension space).

The SAM classification resulted in the detection of eighteen image endmembers, each with different areas mapped as potentially homogeneous regions of *Phragmites*. Visual examination of the eighteen endmembers involved determining if mapped areas generally coincided with areas of *Phragmites* observed in black-and-white aerial photos (1999) and field data collections (2001). Additional validation of mapped areas of *Phragmites* was also aided by using the ENVI Mixture Tuned Matched Filtering (MTMF) algorithms. Visual interpretation of the MTMF *infeasibility values* (noise sigma units) versus *matched filtering values* (relative match to spectrum) further aided in the elimination of potential endmembers. The matched filtering values provided a means of estimating the relative degree of match to the *Phragmites* patch reference spectrum and the approximate subpixel abundance. Correctly mapped pixels had a matched filter score above the background distribution and a low infeasibility value. Pixels with a high matched filter result and high infeasibility were *false-positive* pixels that did not match the *Phragmites* target. At the end of the endmember selection process three *Phragmites* maps were created, one from the northernmost stand and two from the southernmost stand (Figure 4.5, inset, arrow). For the purposes of determining adequate sized areas of mapped *Phragmites*, the three endmember maps were combined as a polygon theme, with a minimum area threshold of 75 m² (i.e., 3 pixels), using ArcView™.

Wetland Field Sampling

Definitions corresponding to the wetland field sampling protocol are listed in Table 4.1.

Vegetation was sampled at Pointe Mouillee on August 7–8, 2001. Prior to vegetation sampling, aerial photographs from 1999 were used along with on-site assessments to locate large target species stands. Six stands were sampled including two stands of each target species and two nontarget vegetation stands for comparison to target-species stands. Digital video of each vegetation stand was recorded to fully characterize the site and for reference during image processing. Each vegetation stand was mapped by a field sketch, noting the general location and shape of vegetation stands, key landmarks that might be recognizable in the remote sensor images, and other information about the site that might be useful when trying to reconcile ground data with remotely sensed data. Transects along the edges of target, *Phragmites*,

TABLE 4.1

Terms Used during Field Sampling Protocol at Pointe Mouillee Wetland

Term	Definition(s)
Wetland	Transitional land between terrestrial and aquatic ecosystems where the water table is usually at or near the surface, land that is covered by shallow water, or an area that supports hydrophytes, hydric soil, or shallow water at some time during the growing season (after Cowardin et al. 1979)
Target plant species	*Phragmites australis, Typha* spp., or *Lythrum salicaria* (per Voss 1972, 1985)
Nontarget plant species	Any herbaceous vegetation other than target plant species
Vegetation stand	A relatively homogeneous area of target plant species with a minimum approximate size of 0.8 ha
Edge of vegetation stand	Transition point where the percent canopy cover ratio of target:nontarget species is 50:50

FIGURE 4.6
Field sampling activities were an important part of calibrating the hyperspectral data. (a) Dense *Phragmites* canopy and (b) dense *Phragmites* understory layer in the northernmost stand. The edges of the stand and the internal transects were mapped using a real-time-corrected global positioning system.

stands (Figure 4.6) were recorded using a real-time-corrected GPS for sampled target species. Each of the two nontarget stands of vegetation were delineated with a minimum of four GPS points evenly spaced around the perimeter. Five GPS ground control points were collected at Pointe Mouillee, triangulating on sampled areas at that wetland. GPS location points were recorded with either a single digital photograph (edge quadrats and nontarget vegetation stand) or multiple digital photographs (ground control points) to provide several angles of each sample location. A written description of each ground control point was recorded to assist in the georeferencing of the remote sensor images.

Within each target-species stand a nested quadrat sampling method was used to sample herbaceous plants, shrubs, tree species, and other characteristics of target-species stands (Mueller-Dombois and Ellenberg 1974; Barbour et al. 1987). Depending on the size of the stands, twelve to twenty (nested) 1.0 m² and 3.0 m² quadrats were evenly spaced along intersecting transects (Figure 4.7). The approximate percent cover and taxonomic identity of trees and shrubs within a 15 m radius was also recorded at each quadrat. Depending on the size of the stand, a transect might either cross the entire stand or penetrate deep into the stand of vegetation. Thus, where appropriate, the terminal quadrat was placed outside of the target-species stand perimeter to characterize the immediately adjacent land cover. The perimeter of each stand and identified corner of each 1 m² quadrat was recorded with a GPS. All GPS locations were recorded with a real-time-corrected (OmniSTAR USA, Inc., Houston, Texas) GPS (Trimble Navigation Ltd., Sunnyvale, California), with a nominal spatial accuracy of 1.0 m. Nonspectral data collected along transects in the canopy and understory vegetation (Figure 4.8) are listed in Table 4.2. Other field spectra, collected with a field spectroradiometer, were considered and analyzed. However, due to extreme spatial and temporal variability in the reflectance values collected for this study, field spectra were not utilized for calibration of sensor data. Caution should always be

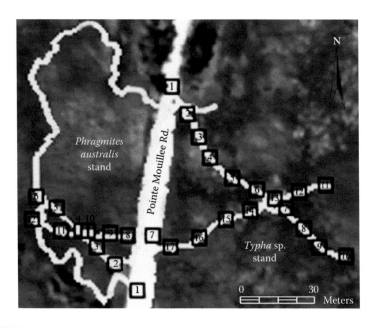

FIGURE 4.7

Magnified view of northernmost field-sampled vegetation stands to the east and west of Pointe Mouillee Road. Two methods were used to quadrat-sample vegetation stands: (1) edge and interior was sampled if the stand was small enough to be completely traversed (left, *Phragmites*); or (2) solely the interior was sampled if the stand was too large to be completely traversed (right, *Typha*). This example shows a *Typha* stand that extended approximately 0.75 km east of Pointe Mouillee Road. Thus the field crew penetrated into the stand but did not completely traverse the stand. Black squares are nested quadrat sample locations. The image is a grayscale reproduction of a natural color spatial subset of an airborne ADAR data (Positive Systems, Inc., Whitefish, Minnesota), acquired August 14, 2001.

exercised when utilizing field-based spectra data for sensor calibration, particularly data from other studies. Field spectra may be archived in a wetland plant spectral library for instructive and other purposes as appropriate.

Results

Vegetation Stand Ecological Characteristics

The northernmost *Phragmites* stand sampled at Pointe Mouillee was bounded on the eastern edge by an unpaved road, with two patches of trees/shrubs to the north (dogwood and willow) and to the south (willow). The eastern edge of the stand was bounded by a mixture of *Lythrum salicaria* and *Typha* spp. Soil in the *Phragmites* stand was dry and varied across the stand from clayey-sand to sandy-clay to a mixture of gravel and sandy clay near the road. Litter cover was a constant 100% across the sampled stand. Nontarget plants in the understory included smartweed (*Polygonum* spp.), jewel weed (*Impatiens* spp.), mint (*Mentha* spp.), Canada

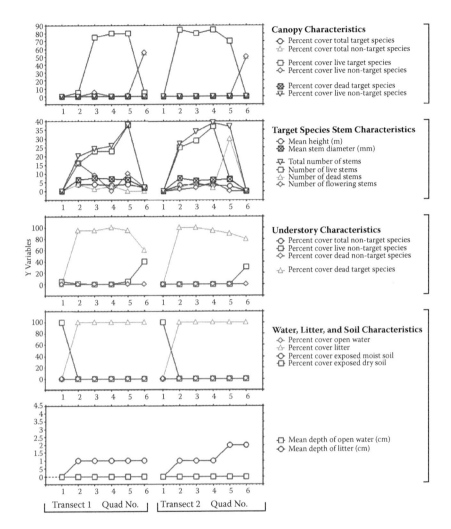

FIGURE 4.8
The heterogeneity of canopy, stem, understory, water, litter, and soil characteristics in
Phragmites stands was used to calibrate the PROBE-1TM data, for the purpose of detecting
relatively homogeneous areas at Pointe Mouillee wetland complex (field data from the northernmost *Phragmites* stand). The most, relatively, homogeneous area of *Phragmites* in the northernmost stand is in the vicinity of transect-1, quadrat-4. Pixels in the vicinity of transect-1,
quadrat-4 were used in the Spectral Angle Mapper (supervised) classification of PROBE-1™
reflectance data.

thistle (*Cirsium arvense* L.), and an unidentifiable grass. Cattail was the sole
additional plant species in the *Phragmites* canopy. Thus the northernmost
Phragmites stand was relatively heterogeneous, with quadrat-4 located in
the most homogeneous region of *Phragmites*, based on the ecological characteristics of the stand.

TABLE 4.2

Nonspectral Data Parameters Collected (√) along Vegetation Sampling Transects at Pointe Mouillee Wetland

Parameter Description	1 m² Quadrat	3 m² Quadrat
Number of live target species stems	√	
Number of senescent target species stems	√	
Number of flowering target species stems	√	
Water depth	√	
Litter depth	√	
Mean stem diameter ($n = 5$)	√	
Percent cover live target species in canopy		√
Percent cover senescent target species in canopy		√
Percent cover live nontarget species in canopy		√
Percent cover senescent nontarget species in canopy		√
Percent cover live nontarget species in understory		√
Percent cover senescent nontarget species in understory		√
Percent cover senescent target species in understory (i.e., senescent material that is not litter)		√
Percent cover exposed moist soil		√
Percent cover exposed dry soil		√
Percent cover litter		√
Percent cover water		√
General dominant substrate type (i.e., sand, silt, or clay)		√
Distance to woody shrubs or trees within 15 m		√
Direction to woody shrubs or trees within 15 m		√
Total canopy cover (area) of woody shrubs or trees within 15 m		√

The southernmost Pointe Mouillee *Phragmites* stand was completely bounded by mowed fields of grass and other herbaceous vegetation. Soil in this stand was dry and clayey throughout. Percent litter cover was 100% and nontarget plants in the understory included smartweed, mint, purple loosestrife, and other grasses. Nontarget plants were not observed in the canopy. Thus the southernmost *Phragmites* stand was more homogeneous than the northernmost stand, with quadrat-8 (not shown) located in the most homogeneous area, based on the ecological characteristics of this stand.

As described, the locations of the most homogeneous regions of *Phragmites* within sampled vegetation stands were determined by examining field transect data to determine which had the greatest percent cover of nonflowering live plants and greatest stem density (Table 4.2). *Phragmites* is a facultative-wetland plant and usually occurs in wetlands, but occasionally occurs in nonwetlands (Reed 1988). Thus it can grow in clayey soil, varying from moist to dry substrate conditions. We did not have a *Phragmites* field sample in a moist-soil area at this site, but considering the great density of vegetation in the canopy and the high stem density, spectral endmember maps of field samples were sufficient to detect relatively homogeneous areas of *Phragmites* at a large number of locations

Phragmites *Spectral Characteristics and Comparisons*

Phragmites and *Typha* spp. are often present within the same wetland and frequently intergraded making them difficult to distinguish from a distance, either on the ground or in remote sensing imagery. Thus we compared the reflectance spectra within a single stand of *Phragmites* (Figure 4.9) and between the two taxa (Figure 4.10) using PROBE-1 data. There was substantial variability of reflectance spectra among pixels within the northernmost stand of *Phragmites* along field-sampled transects. The greatest variability of *Phragmites* along the two sample transects, within the spectral subset 470–850 nm, where reflectance from photosynthetic plant material was the greatest, occurred in the 740–840 nm (near-infrared) wavelength range. Comparison of reflectance characteristics between *Phragmites* and *Typha* spp., using the nearest pixel to *Phragmites* quadrat-4 and *Typha* quadrat-8 (i.e., the most homogeneous area in each stand in Figure 4.7), indicated that *Phragmites* reflected less energy than *Typha* in the near-infrared wavelengths and reflected more energy than *Typha* spp. in the visible wavelengths.

Semiautomated Detection of Homogeneous Phragmites *Stands*

Analyses of field measurement data (Figure 4.8), digital still photographs, digital video images, field sketches, and field notes resulted in the selection of nine relatively pure pixels of *Phragmites* centered on quadrat-4 (Figure 4.7). SAM-supervised classification of the Pointe Mouillee PROBE-1 image resulted in a vegetation map indicating the locations of other relatively homogeneous *Phragmites* stands (Figure 4.11). Several of these areas are located in the diked areas of the wetland complex, areas that were typically populated by large stands of *Phragmites* in other Lake Erie wetlands.

Accuracy Assessment of Semiautomated Phragmites *Vegetation Maps*

A three-tiered approach to accuracy assessment of semiautomated vegetation maps at Pointe Mouillee was followed at the remaining twelve wetland study sites. The accuracy assessment approach included (1) testing of target plant species presence or absence using a comparison of semiautomated vegetation maps to recent stereo aerial photographs; (2) testing of target plant species presence or absence using random field samples of the mapped areas;

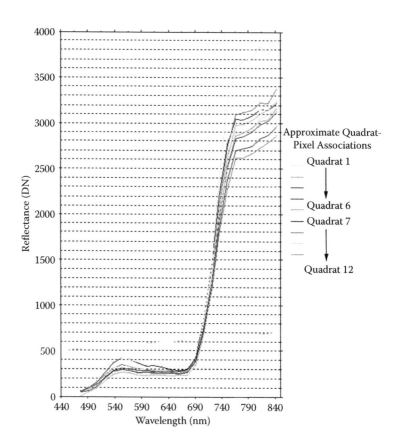

FIGURE 4.9
Demonstration of variability in the spectral reflectance of vegetation along each of the 2 (approximately 60 meter) transects in the northernmost sampled stand of *Phragmites australis* at Pointe Mouillee. The greatest variability in reflectance occurred in the near-infrared bands. Reflectance data are derived from a 470 nm–850 nm spectral subset of PROBE-1™ data, acquired in August, 2001.

and (3) testing of target plant species percent cover and structural composition using random field samples of mapped areas.

At Pointe Mouillee, tier-1 accuracy assessment (prior to field validation sampling) compared vegetation maps to 1:15,840 scale, black-and-white stereo aerial photographs (September 1999) and field notes (May–August 2001). Tier-1 accuracy assessment results indicated that approximately 80% of the areas mapped as *Phragmites* were located within true *Phragmites* stands. Field sampling to complete tier-2 and tier-3 accuracy assessment was performed in August 2002. Comparison of field samples with the semiautomated vegetation maps of *Phragmites* resulted in a 91% user accuracy ($n = 86$). Tier-2 accuracy assessment of *Typha* spp. and *Lythrum salicaria* at the other wetland study sites and tier-3 accuracy assessment at Pointe Mouillee found similar results.

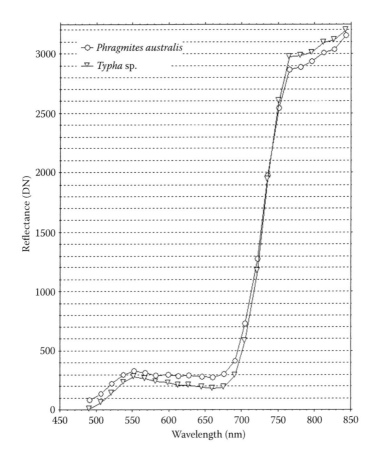

FIGURE 4.10
Comparison of *Phragmites australis* and *Typha* spp. spectral reflectance in a relatively homogeneous region of each stand (sample size = 5.0 m × 5.0 m) using PROBE-1TM data (470 nm–850 nm spectral subset). The location of pixels are in the northernmost field-sampled *Phragmites* stand quadrat-4 and *Typha* stand quadrat-8.

Discussion and Conclusion

Because the ADAR data had a nominal spatial accuracy of 75 cm, it was a useful and convenient data (i.e., digital) format for viewing field GPS overlays, similar to color infrared aerial photographs. However, because ADAR is 4-band multispectral data it was limited in its usefulness for developing spectral signatures of *Phragmites*. Nevertheless the relatively small ADAR data file size (approximately 5 Mb per 1.1 km × 0.7 km scene) allowed for relatively efficient georeferencing using image-to-image warping techniques.

Field data from quadrat sampling was an essential part of effectively using PROBE-1 data to map *Phragmites* at Pointe Mouillee. The nominal 1.0 m

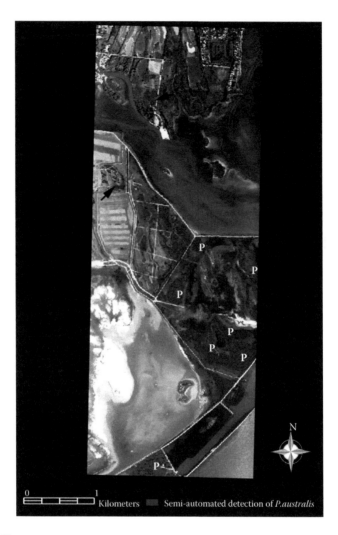

FIGURE 4.11
(See color insert.) Results of a Spectral Angle Mapper (supervised) classification indicating likely areas of relatively homogeneous stands of *Phragmites australis* (solid blue), using PROBE-1 data and field-based ecological data. Black arrows show field-sampled patches of *Phragmites*. Areas of mapped *Phragmites* are overlaid on a natural-color image of Pointe Mouillee wetland complex (August 2001). The letter P indicates the general location of known areas of *Phragmites*, validated with aerial photographs, field notes, and 2002 accuracy assessment data.

spatial accuracy of the GPS data, by way of real-time correction, and the eco-logical data helped to demonstrate the heterogeneity that exists in *monospe-cific stands* of *Phragmites*, resulting from variability in underlying vegetation, litter, and soil conditions. Thus, the use of georeferenced field data enabled us to select specific pixels within the PROBE-1 image that contained the most homogeneous regions of *Phragmites* (Figure 4.11). Additionally, ground imagery (i.e., video and digital still images) of each quadrat improved the decision-making processes about which regions of the vegetation stand were dominated by live, nonflowering *Phragmites*.

Field results at Pointe Mouillee demonstrated that a major impediment to semiautomated detection of wetland vegetation was the heterogeneity of biotic and abiotic characteristics within plant communities, even when the vegetation was dominated by a single taxa, such as *Phragmites*. Although water was not present at the selected Pointe Mouillee 2001 sample locations, the presence of water and variable soil moisture could be a confounding factor in many wetlands. Thus hydrologic variability is a potential confounding variable when attempting to identify or match spectral characteristics among *Phragmites* stands.

At Pointe Mouillee, we minimized the potential remote sensing pitfalls of plant community heterogeneity by (1) selecting plant taxa that were least likely (by definition) to exist in diverse, heterogeneous plant communities; (2) using GPS points with a nominal spatial accuracy that exceeded that of the acquired remote sensing data for locating sampled quadrats, edges, and ground control points; (3) acquiring and using a variety of remote sensing data types that had a range of spectral or spatial characteristics for detecting wetland plants; (4) collecting and using relevant ecological field data that were most likely to explain the differences in spectral reflectance characteristics among pixels; (5) collecting and using an abundance of ecological field data to sufficiently account for phenological variability and variability among species, quadrats, stands, and wetlands; (6) collecting and using historical remote sensing data for contextual information about the site; and (7) collaborating with local wetland experts to better understand the ecological processes of the sites, in a historical context.

PROBE-1 data demonstrated that a *relatively pure* pixel of *Phragmites* was quantitatively different from a *relatively pure* pixel of *Typha*, which may be the result of differences in areal coverage, chlorophyll content, physical structure, or water relations between the two taxa. Although *Phragmites* and *Typha* are both tall monocots, their basic physical structure is quite different in that *Typha* is primarily comprised of photosynthetic *shoots* that emerge from the base of the plant (at the soil surface); *Phragmites* has a main stem that is fibrous, with branching leaves. *Phragmites* also has a large seed head that varies in color from a reddish brown to a brownish black; *Typha* has a relatively small, dense, cylindrical seed head. Thus physical differences in plant structure in combination with soil conditions, hydrologic conditions, and mixed vegetation conditions may contribute to the spectral differences observed between *Phragmites* and *Typha*.

Summary

The combined use of ecological field data and SAM-classified PROBE-1 data demonstrated how dense stands of opportunistic plant species, such as *Phragmites australis*, could be mapped over large wetland areas. The vegetation map produced at Point Mouillee solely addresses the purest areas of *Phragmites*, that is, was an estimate of the most homogeneous areas of *Phragmites*. Thus there were other areas of *Phragmites*, mixed with other plant species of lesser density (i.e., percent cover), surrounding the mapped area. Because these classification results were specific to this PROBE-1 scene, results may not be entirely transferable to other locations or to imagery acquired during another time period.

These results are among the first steps toward investigating the landscape–ecological relationships between the extent and pattern of opportunistic plant species in wetlands and local landscape disturbance. These results build upon ongoing efforts in the Great Lakes to develop new vegetation mapping methods and may lead to innovative techniques for identifying additional plant taxa and communities in wetlands or in other ecosystems. The airborne hyperspectral approach can be used to good effect in the detection of dense patches of *Phragmites australis*, a native opportunist plant species. This approach provides a semiautomated process of mapping the presence of invasive plants like *Phragmites australis*.

The results of this case study and the techniques outlined in this chapter provides the practitioner with the tools and procedures for integrating field-based and geospatial data to successfully accomplish wetland landscape characterization projects and programs. Other chapters provide the broader scale context to apply those successes to regional issues and other challenges.

5

Utilizing Broad-Scale
Characterization Approaches

Questions about broad-scale wetland conditions are an important part of framing the dialogue. These diverse systems exist across the landscape and are often mixtures of conditions (Figure 5.1). Questions include: How much do we have? Are they in danger of degradation? For purposes of addressing important issues, the public often needs to know what these trends are across a vast area, regionally or nationally. The public continually poses these questions and they are the main examples asked by decision makers and managers as well. Hence it is vital to obtain data that can establish fundamental knowledge as a basis for charting a vision that includes a vibrant place for wetlands. In other words, one needs to know what one has and what the opportunities are to increase wetland resources.

Historical work has focused on combining statistics from agricultural, natural resource, and conservation inventories. The reports often talk of historical numbers at the time of the U.S. Public Lands Survey or as homesteaders moved lands into agricultural or range production. Such inventory statistics had been kept in support of agriculture programs and, for example, include acreage of wetlands drained and placed into production. These assessments are useful for establishing the status of wetlands at a given time, and in some cases for trends in amounts of total wetlands. They tend to be broad in definition and in area of coverage (Dahl and Johnson 1991; Dahl 2006; Dahl and Watmough 2007).

These historical inventories seldom provide maplike details and mapping accuracy or precision characteristics (Van Genderen and Lock 1977; Congalton and Green 1998, 2009). They were state of the art in their time and the best that could be done. Naturally, many have hoped for a national or regional inventory with such characteristics. The U.S. Fish and Wildlife Service (USFWS) National Wetlands Inventory (NWI 2012; www.fws.gov/wetlands/; accessed July 21, 2012) is such an attempt and supplies good information and mapping products. The value and limitations of the NWI are discussed throughout the text, and addressed in detail in Lyon and Lyon (2011). There is a continued need for maplike inventories in support of the dialogue and to do so with known accuracy and precision, and in the context of user needs and societal goals.

These maplike inventories and remote sensor data have been contemplated on a number of scales, and details are presented here both as to methods and

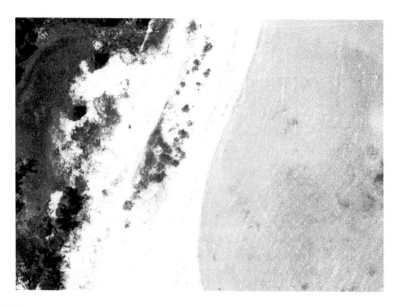

FIGURE 5.1
Differences in perspective, scale, sensors or emulsions, and time of day or year all can give clues to the characteristics of wetlands. This series of images show how one can develop a convergence of evidence from several images to arrive at a conclusion. This low-altitude image illustrates one data point in the convergence and shows how tone, texture, and pattern can reveal characteristics. Pictured here are the dark-toned individual plants on the surface of the light-toned sandy beach barrier. Note the dark, fairly uniform tone of shrub cover uphill and to the left of the image. Note that trees (versus shrubs) have a shadow that helps segregate them from the similar dark-toned shrubs.

outcomes. The results have been possible because they relied on a definition of wetlands different from that of jurisdictional wetlands (Anderson et al. 1976; Cowardin et al. 1979; Brown et al. 1979; Omernik 1987, 2003). This is because delineation of jurisdictional wetlands (JW) requires field work, which is often done on a limited basis for inventories. Hence there is a need for a definition of wetlands of a general nature or potential JW that allows gathering of inventory-scale information.

General wetlands or potential jurisdictional wetlands (PJW) are defined as possessing one, two, or three of the wetland criterion (Lyon 1993; Lyon and Lyon 2011). Those criterion characteristics have been agreed upon by any number of experts or expert bodies, and include an abundance or dominance of wetland-loving plants, wetland hydrology, and wetland or waterlogged soils (Cowardin et al. 1979). By using a definition of general wetland areas possessing one, two, or three of the criterion indicators, we can map or inventory from a distance (Figure 5.2). Conversely, limited field work can be supported to develop the inventory or validate the accuracy and precision of the inventory numbers (Congalton and Green 2009; O'Hara et al. 2010).

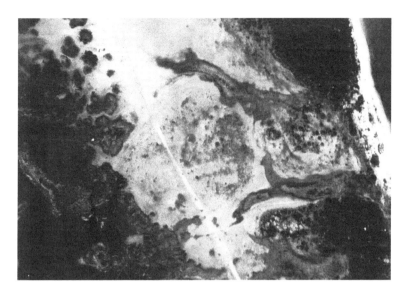

FIGURE 5.2
Another image in the series from higher altitude includes the area depicted in Figure 5.1, to the right of this image. Tone, texture, and pattern can again tell a story. The center of the image is a mix of light and dark tonal pattern, which is the expression of grasses and shrubs obscuring the light-tone sand and pebble outwash flats. The path from top to bottom has very little vegetation on it and allows one to see the pure sand spectra or reflectance without dark toned plants. To the right, the mixture of grays and darker gray tones reveals the abundance of wetland grasses, sedges, and also *Scirpus* species in the washover channels from the straits side of the Straits of Mackinac. To the left, dark gray-toned plants in dense canopy obscure any light tone reflectance of the sandy sediment beneath the wetland *Juncus* canopy. Although it is difficult to identify species from imagery, one can identify plant communities by knowledge of the area and field visits along with plant community mapping from imagery, as described in Chapter 4.

There are several methods to develop historical numbers of general wetlands or potential jurisdictional wetlands (Lyon 1993; Lyon and Lyon 2011). The sources of information can include old maps, survey records, and governmental records (General Land Office Records, www.glorecords.blm.gov/ or www.blm.gov/wo/st/en.html; accessed July 21, 2012) as well as inventories mentioned here (e.g., for Wisconsin http://dnr.wi.gov/topic/Wetlands/ inventory.html; accessed July 21, 2012). Naturally, these are subject to a certain potential for errors, either known or unknown. It is also difficult to reconcile the definition of a given wetland resource in the history of resource inventories, as compared to the definition used in the current timeframe as an outcome for a given project.

In addition to the challenges of wetland definition and obtaining wetland data, there is the variability of the resource itself. Lake and marine coastal wetland ecosystems exhibit changing characteristics due to the variability of water and terrestrial resources. This variability is an important characteristic, and can supply great wetland functions such as fertility due to sediment deposition, or changing soil chemistry that can process or sequester nutrients and release other nutrients for productivity. This variability is also a challenge to their measurement (Verbyla and Hammond 1995).

The variability is repeated in most landscapes. The lake system experiences an influx of water from runoff, groundwater, and interception of precipitation. The origin of the lake and the characteristics of the watershed largely govern lake water quantity and quality issues, and govern to a certain extent the quantity and variety of wetlands. Likewise, marine coastal wetland areas are defined by water resource characteristics and range from freshwater coastal lakes to marine coasts to the coasts of hypersaline lakes or seas often found in arid settings. Lakes, rivers, and coastal areas are also greatly influenced by their three-dimensional characteristics, such as water depth, and the relation between water depth and forcing functions or stressors such as river flow and flooding, wind-driven currents, thermal stratification, and evaporation and transpiration issues in shallow lakes (Figure 5.3).

Coastal wetland areas exhibit many unifying characteristics and some greatly differing characteristics. By definition coastal areas are shallow in water depth as compared to the depth of the open lake, sea, or ocean. Many of the techniques used to characterize coastal areas are the same as those

FIGURE 5.3
A low-altitude image from the series shows the plant canopy and tree crown characteristics of the forested wetlands and shallow lake marshes along the shore. Tone, pattern, and texture demonstrate very little if any submergent vegetation that would be dark toned against the very light-toned marl lake bottom. The lake edge marsh wetlands are easily separated from the marl bottom by their dark tone and varying texture, and easily separated from the forested wetland by the crown shapes and shadows.

used in lake and wetland areas (Lyon 1993; Lyon and Lyon 2011; Wang 2010). The interplay of tides and wind-driven events such as flooding further complicate the picture.

The methods used to evaluate wetland characteristics and extent, and supply data for remote sensing and GIS analyses are often set by the goals of the project and by the scale of the effort. Although previous efforts supply regional and broad knowledge, remote-sensor-based methods supply detailed wetland types and spatial location. This specific knowledge of wetlands is necessary and is required for today's dialogue, particularly with regard to analyses at a variety of scales.

Temporal Change

Evaluations of wetlands and their change over time have been completed using aerial photographic or image products and interpretations (Lyon 1993; Lyon 2003; Adam et al. 2010; Lyon and Lyon 2011). Evaluations of multiple dates of historical images allow for interpretation of conditions and changes over time (Garofalo 2003). It is recommended that one use four or more dates of images, if possible, to look at various features over time (Lyon and Lyon 2011). The abundance of imagery in archives, currently acquired imagery, and methods to integrate these data for analyses make historic or inventory projects attainable (Figure 5.4).

From a number of dates of historic photographs or images, hydrological, vegetation, soils, and human activity data can be interpreted. It is possible to evaluate the presence or absence of hydrological conditions or stressors and their exposure to general wetland areas. These efforts have been described for isolated wetlands in Chapter 3 and elsewhere (Federal Geographic Data Committee 1992, 2008, 2010; Lyon 1993; Lyon and Lyon 2011).

Many current efforts have made use of remote sensor products for analyses of change (Lee and Marsh 1995; Schaal 1995; Lunetta and Elvidge 1998; Garofalo 2003; Vicente-Serrano et al. 2008; Lunetta et al. 2010). There are many opportunities for digital remote sensor analysis efforts, as this approach allows for more specific identification of wetland types based on spectral characteristics in the visible, near and middle infrared, and microwave portions (Figure 5.5) of the electromagnetic spectrum (Lyon and McCarthy 1995; Pollard et al. 2010).

Another advantage of remote sensor methods is the utility of image processing of data and modeling of data using GIS (e.g., Maidment and Djokic 2000; Maidment 2002; Lyon 2003). The advent of image processing has added great analytical power to evaluations including detection of change in wetlands. Further, the availability of fine-resolution satellite data with very good repeat coverage has brought a whole new dynamic to the effort (Office of

FIGURE 5.4
This next image in the series can be interpreted using the same characteristics even though the scale may blur some features. The mid-image forested wetlands can be separated from the foreground shrub wetlands by canopy and crown shape and tone, and the more uniform texture of the shrub scrub wetlands.

Science and Technology Policy 2010). These approaches are discussed as they supply a great deal of capabilities to the dialogue (Figure 5.6).

Efforts have been made to develop methods for efficient and accurate evaluations of land cover change from airborne and satellite remote sensor data (Adam et al. 2010; Klemas 2011). These procedures were initially developed on wetland ecosystems in North and South America including the trans-Amazon as well as Africa for purposes of evaluating climate change (Thenkabail et al. 2005; Velpuri et al. 2009). Due to the importance of change detection in research and management missions, large efforts have been devoted to the development of these procedures for operational evaluations of change from satellites (Ramsey and Jensen 1995; Ramsey et al. 1998; Thenkabail et al. 2012). Currently, use of twice daily MODIS coverage advances this endeavor and adds greater potential for monitoring phenomena and phenology as influencing wetland characteristics. This application represents a high development of the art and supplies great utility to the user much like Google Maps does (Linderman et al. 2010; Lunetta et al. 2010). Other related applications of twice daily, moderate-resolution data have demonstrated the utility of these resources.

Early projects focused on climate change issues related to forested areas in the Pacific Northwest, Africa, Asia, and the trans-Amazon. Once methods were developed the work moved on to more regional studies, such as change detection of land covers in all of Mexico crossing three decades as part of the

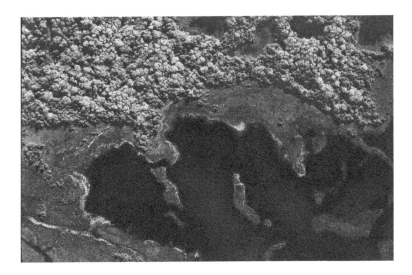

FIGURE 5.5
Wetlands are a mixture of water, soils, and plants, and the combination can be identified often by interpretation of characteristics. Here, in this series image sand or sediments are very light toned, and shallow water depths only slightly gray the tone. Deeper water extinguishes the light that could have entered the water, gone to bottom depth, and been reflected and upwelled to the surface. Hence, deeper waters or optically deep waters are very dark toned. At the top of the image, plants are very dark toned too, much like the optically deep water, but have texture related to crown shape and shadow, and can be distinguished from equally dark-toned water that have little or no texture.

North American Landscape Characterization (NALC) project (Lunetta et al. 1993; Lyon et al. 1998; Lunetta et al. 1998; Vicente-Serrano et al. 2008).

Change detection methods have been broadly divided into either enhancement (precategorization) or postcategorization methods (Thenkabail et al. 2009; Sen et al. 2012; Sun et al. 2012). Either approach can have utility in wetland change detection. Selection of a given approach may have roots in scale or spatial resolution issues, or in spectral resolution.

Image enhancement change detection techniques involve the transformation of two original images to a new single-band or multiband uncategorized image in which the areas of land cover change can be detected. The resulting image data can be further processed by other methods, such as by using a categorizing algorithm, to produce a categorized change detection product or thematic map of change, such as in the Texas case study in Chapter 3.

These enhancement techniques or preclassification techniques were often based on the concepts of image differencing or image ratioing (Schott 2007). Differencing of vegetation indices has proven to be a valuable approach for effective detection of change in sensor images (Lyon et al. 1998; Thenkabail et al. 2009; Thenkabail et al. 2012).

Image equalization is valuable in the data preprocessing stage as it usually improves results of change detection (Lunetta and Elvidge 1998). Techniques

FIGURE 5.6
Use of other parts of the spectrum such as the infrared can add more details for the convergence of evidence. This color infrared image rendered in black and white displays the bright tone in the near-infrared of healthy, vigorously growing marsh wetland vegetation adjacent to the trees in the center of the image, as well as the same vegetation but with standing, dark-toned waters intermixed with the vegetation.

like band-to-band regressions and principal components analysis have been used to simultaneously perform the image-to-image equalization and the detection of change areas.

In postcategorization change detection, two images from different dates are independently categorized. The area of change is then extracted through the direct comparison of the categorization results. A key advantage of post-categorization change detection is that it bypasses the difficulties in change detection associated with the analysis of images acquired at different times of year or by different sensors. The disadvantages of the postcategorization approach include greater computational and labeling requirements, high sensitivity to the individual categorization accuracies, plus the difficulties in performing adequate accuracy assessment on historic data sets (Lunetta et al. 1993; Sinha et al. 2012).

Accuracy in image registration and coregistration is critical because all enhancement methods are based on pixelwise operations or scenewise plus pixelwise operations. This is true even with moderate-resolution satellite data such as MODIS sensors on Terra and Aqua NASA satellites, where spectral signatures are developed over wide areas of coverage (Thenkabail et al. 2005; Johansen et al. 2010; Sun et al. 2012).

Sensor products can be obtained and processed by the user into enhanced products of interest. These enhanced images could include a greenness or biomass index image for each date of coverage, other ratio type images, and unsupervised (or supervised, as in the Michigan case study described in Chapter 4), categorization images with approximately one hundred classes. As part of the project efforts, a number of methods have been identified to enhance the data sets and methods to conduct change detection experiments. Much of the change detection and other experiential results can be further explored in the literature, and is well evaluated by Lunetta and Elvidge (1998) and Thenkabail et al. (2009).

The National Land Cover Data (NLCD) carried forward the ground-breaking work of North American Landscape Characterization (NALC) project and Multi-Resolution Land Characteristics (MRLC) consortium, and provided similar Landsat-based, 30 m resolution land cover information for the United States. The NLCD provided spatial reference and other important descriptive data for the characteristics of land surfaces, which include urban, agriculture, forest, percent impervious surface, and percent tree canopy cover. It has supported numerous federal, state, local, and non-governmental programs and specific applications, and provided the public with the opportunity to evaluate ecosystem status and condition, improving the capability of users to (1) understand spatial patterns of biological diversity; (2) address climate change and associated outcomes; (3) look at ecosystem/ecological services and land health; and (4) develop land management policies and procedures.

NLCD products were developed and produced by the MRLC Consortium, which has continued as a multidisciplinary partnership of federal agencies led by the U.S. Geological Survey (USGS). All NLCD data products are now freely available and downloadable from the MRLC website (www.mrlc.gov; accessed July 21, 2012), and products can be viewed on the MRLC consortium viewer at the same site.

A measure of the value of MRLC products was the second set of content developed in the last decade referred to as MRLC 2006 (Vogelmann et al. 2001; Foody 2002). The net result of this leadership over four decades is the availability at no cost of land cover data over several decades and at several spatial resolutions for the United States. The user can utilize these data sets to develop regional or local analyses of general wetlands. The availability of the Landsat data archive at no cost from the USGS allows for deeper analyses or temporal studies. This is a great boon to the new users of land cover data as well as the more capable user who has varying familiarity with remote sensed data (Na et al. 2010).

Other worthwhile remote sensing products can be used for change detection or as ancillary data (Barrette et al. 2000; Abood et al. 2012). They can include the U.S. Department of Agriculture's National Agriculture Imagery Program (NAIP) and High Resolution Orthoimagery (HRO), which can be downloaded free of charge at the USGS's Seamless Data Warehouse Internet website (http://nationalmap.gov/viewer.html; accessed July 21, 2012).

Data Processing

There are a number of approaches and methods for developing remote sensor data for detection change in wetlands over time. Often users wish to develop their own data sets from basic remote sensor data as opposed to using available land cover data such as MRLC. The process begins with selecting multiple date data sets to identify the wetland or other features of interest, and assembling them to simplify the detection of change using image processing.

Characteristics of sensor data need to be addressed to achieve successful results. The spatial and spectral characteristics should be similar to facilitate analyses (Vicente-Serrano et al. 2008). If the data sets are from different sensors or bands in the electromagnetic spectrum, the resolution and spectral bandpass differences between images acquired with two or more sensors complicate the direct comparison or the digital analysis of the data to detect change. These issues but can be addressed through advanced technology and its thoughtful application.

With the spectral bandpass differences, the detectable land cover classes for the two dates of imagery may not be comparable. Land cover classes that are distinct when observed with one sensor may be indistinguishable with a sensor having broader bands or fewer numbers of spectral bands. If there are substantial spatial resolution differences between the two input images, ground features may be visible in one data set and undetectable in the other.

Variations in the radiometric response of a sensor can complicate change detection by requiring some form of image-to-image equalization prior to the change detection or by requiring the use of a postclassification change detection approach (Vogelmann et al. 2001). Again, this can be addressed as it was in the Texas case study described in Chapter 3.

If clouds are present in images from one or both dates, it can be impossible to detect land cover change between selections of two dates of imagery where clouds occur. To minimize the effects of clouds, low cloud cover imagery is typically a primary factor in the selection of scenes for change detection efforts. In addition to clouds obscuring the land below, clouds also shadow other land areas reducing the total solar insolation and subsequent reflectance or darkening the look of the shaded area. This is also a challenge as we used differential reflectance to characterize materials, and a shadowed area of a given land covers will appear different than the identical land cover in direct sunlight.

Variations in solar irradiance, solar zenith angle, and solar azimuth will affect scene brightness levels and the location of shadows. The change detection projects have attempted to reduce these effects by selecting scenes for the three time periods that were acquired at or near the same date in order to match the solar conditions as far as possible (Schott 2007). In general, high sun angles (low amount of shadowing) are better than low sun angles for the detection of land cover change.

Variations in atmospheric effects (scattering and absorption) can affect scene characteristics sufficiently that they must be considered in evaluating change detection methods (Lunetta and Elvidge 1998; Schowengerdt 2007). The change detection data sets have been visually screened to remove scenes where within-scene atmospheric variations are obvious, and now the use of *browse* files composed of metadata and compressed images can be used to select optimal images. By assuming that the atmospheric effects on the selected scenes are uniform across the entire scene area, it is possible to partially compensate for scene-to-scene variations in atmospheric effects by scene-to-scene brightness equalization (Lunetta and Elvidge 1998). The coregistration of the image products will also help to compensate for topographic variations that alter atmospheric path length.

Phenological variations in vegetation result in large changes in the reflectance patterns of the land surface. If images of leaf-on and leaf-off conditions are compared, whole regions can appear to have been *deforested*. The scene selection process of the change detection projects have attempted to minimize this problem by selecting scenes from the same time of year for each of the selected time periods. In some cases it was not possible to do this due to the lack of appropriate archival data for comparison purposes (Fassnacht et al. 2006; Sinha et al. 2012).

Spatial miss-registration of images will tend to reduce the accuracy of any digital change detection effort. These effects are most severe on the change detection techniques using enhancement. For the change detection, project data sets are digitally coregistered, with accuracies on the order of half a pixel or less, creating data sets that can be analyzed for change with minimal errors due to miss-registration (Lunetta et al. 1993).

Vegetation Index Differencing

The development of vegetation indices from spectral reflectance values is based on the differential absorption and reflectance of energy by vegetation in the red and near-infrared portion of the electromagnetic spectrum (Thenkabail et al. 2009; Thenkabail et al. 2012). In general, green vegetation absorbs energy in the red region and is highly reflective in the near-infrared region (Lillesand et al. 2004). A number of vegetation indices have been formulated and utilized for monitoring vegetation change. Of these vegetation indices, the Normalized Difference Vegetation Index, or NDVI, has been used most widely for monitoring terrestrial vegetation dynamics (Lyon et al. 2003). The NDVI compensates for some radiometric differences between images; however, it does not completely remove radiometric images that are being compared. The difference in the NDVI values of two images in certain cases responds to changes in land cover. Singh (1989) concluded that NDVI differencing was among the most accurate of change detection techniques, and many indices have been developed and tested, and they demonstrated great utility (Thenkabail et al. 2009; Thenkabail et al. 2012).

Importance of Preprocessing

One can use a set of procedures that will render high-quality image scenes for derivation of ratio and index-type images, and for categorization of land cover types. The goal of applying these algorithms is to produce images with few system-related sources of noise. All or some of these procedures may apply to a given analysis or application, and the selective use of each is dictated by the type of sensor data and by the objectives of the effort. The procedures include smoothing of spectral variability, image compositing to remove clouds, image enhancements such as ratioing and indices, and others.

Processed images will be often be smoothed or deconvoluted by *destriping* of images. This procedure removes the variability from row to row in the image due to differential response by one or more radiometric problems. Often there will be a general deconvolution of processed images, as it is often desirable to resample the pixel size to a known quantity (say, 57×80 m for MSS data or 30×30 m for TM data) and maintain geometric fidelity. Special efforts were made over the years to develop methods to create cloud-free composites of multiple date images of the same area (Loveland and Ohlen 1993; Chander et al. 2009). This is necessary as some image scenes will be collected and they will exhibit clouds, and much work was conducted by the USGS EROS Data Center to develop these methods in support of the Advanced Very High Resolution Radiometer (AVHRR) biweekly vegetation index composites program over the years (Young and Wang 2001; Vicente-Serrano et al. 2008; Vogelmann et al. 2001), and similar programs using MODIS data (Linderman et al. 2010; Lunetta et al. 2010).

The composite images are made of cloud-free portions of images from different dates, and processing procedures will also eliminate systematic and random variability. This variability results from changes in the sun's position, atmospheric influences, vegetation growth patterns or phenology, and other reasons. This variability causes the same or materials or land cover types to exhibit dissimilar spectral signatures over time.

A portion of this variability is reduced by calculating the Normalized Difference Vegetation Index (NDVI) using the red and near-infrared portions of the electromagnetic spectrum. A processing system can be developed to check the NDVI values of identical position pixels in each image under study. NDVI values that are commonly associated with cloud-obscured pixels can be avoided, and only cloud-free pixel values can be written to the output image.

The composite images often will undergo additional processing to make the component images as spectrally similar as possible. Corrections of the scenes to a set solar zenith angle will establish similar solar illumination conditions (Schott 2007; Chander et al. 2009). This can be affected by the image equalization procedures described in detail by Lunetta and Elvidge (1998) and in others (Thenkabail et al. 2009; Vogelmann et al. 2001).

The NDVI is used to make this valuable composite product, such as the product in Figure 4.2. Many uses have been found for NDVI that illustrate the utility of biomass or greenness indices. For example, the calculation of the index or ratioing of image bands is valuable because it reduces the differential solar illumination effect. This occurs when mountainous areas receive more illumination on the sun-facing slopes than on the slopes facing away from the sun. Another valuable feature is that NDVI also reduces atmospheric haze and other systematic noise contributions. These band ratios may be used as products in themselves or as input to categorization algorithms.

Postcategorization

Postcategorization change detection involves the categorization or computer classification of images, and labeling of land cover thematic classes from each year using the same class types from the same classification systems (Bolstad and Lillesand 1992; Villeneuve 2005; Baker et al. 2006). The locations of change can be identified by the areas of change in land cover from the earlier date image to the later date image.

As in the preclassification change detection methods, the input image data can be optimized to remove sources of error that are spectral and geometric. The aforementioned examples include sensor noise (e.g., stripping), atmospheric effects or haze, and differential solar illumination. After data enhancements, the input images can be used in categorization.

The data are to be processed by steps that include image preprocessing, training set selection, training set evaluation, and maximum likelihood categorization. Image analysts or cooperators can identify the clusters or classes of the preliminary land cover (PLC) maps, aggregate and assign land cover types from the categorization system, and create the land cover (LC) products.

Land Cover and the Role of the Classification System

The land cover classes in the PLC image, previously described can then be labeled as to cover type using a land cover classification system (Jensen 2004). Classification systems were developed to specifically support project objectives. Most classifications can be interpreted with the other major land cover classification systems.

Any custom land cover classification system should be compatible with the Anderson et al. (1976), Cowardin et al. (1979), and other commonly used and accepted systems. These systems have been optimized to inventory applications, which include wetland land cover types. This greatly facilitates understanding by later users.

The assignment of class types to PLC classes should be performed by personnel familiar with these techniques and experienced in this practice. The work may also be done with the help of local cooperators familiar with the land cover and wetland land cover of the study area. Naturally, this sort of work is greatly facilitated by fieldwork and supporting information such as other studies, soil surveys, crop information, topographic, or digital elevation mapping data, or other land cover information such as NALC or NLCD.

The level of detail a classification system can be selected by client groups. Level 2 or level 3 is often used as the final level of land cover type detail to be identified. This approach will supply sufficient detail while using the capabilities of remote sensor data to their best advantage. It is recognized that differentiation between wetland forest types and wetland shrub-scrub classes at level 2 will be difficult and in some cases not possible without the aid of ancillary data such as RADAR or LIDAR or Shuttle topography data (Lunetta et al. 2010; Abood et al. 2012).

The final categorization images should be accompanied by details on the training sets including cluster or class means in each band, variance–covariance matrix, class identities, and details related to quality and accuracy. The resulting products will be a land cover (LC) thematic data coverage product for each scene. It can then be supplied to the client groups for validation. Subsequent to data validation, LC products should be archived for reference and later use.

The land cover classification system used by the NLCD provides a good and recent example. It contains land cover information of a general nature as well as general wetland classes. It provides a model as to how this work is done and can also support more detailed, localized studies.

Advanced Sensors

The continued introduction of advanced sensors has made issues of identifying and characterizing inherently variable wetland resources more facile (Antalovich 2011). The use of multiple parts of the electromagnetic spectrum and fine spatial resolution enhances this work immensely (Gilmore et al. 2010; Klemas 2011).

There are still elements of seasonal coverage, availability of historical sensors of similar capability for change analyses, and spatial resolution challenges. Even these difficult analyses have been addressed using daily coverage by sensors such as MODIS (https://lpdaac.usgs.gov/products/modis_products_table. Accessed 12/9/12) to follow the variability of vegetation and hydrology (Thenkabail et al. 2005; Lindeman et al. 2010; Lunetta et al. 2010; Sun et al. 2012).

A very appreciable capability is that of spaceborne sensors of high spectral fidelity and fine spatial resolution (Klemas 2009). There are many examples and they are increasing in utility such as thermal images, MODIS, ASTER, and other imagery (Thenkabail et al. 2005; Wolter et al. 2005; Callan and Mark 2008; Pantaleoni et al. 2009; Yang et al. 2009; Akins et al. 2010; Rodrigues-Galiano et al. 2012). Hyperspectral sensors are also demonstrating great purpose (Zomer et al. 2009; Thenkabail et al. 2012).

The utility of RADAR or microwave remote sensing is particularly attractive (Lyon and McCarthy 1981; Wu 1989; Ramsey 1998; Ramsey et al. 1999; Islam et al. 2008; Lang et al. 2008). Most RADAR wavelengths respond to dielectric constant or conductivity nature of the materials and to the roughness of the canopy structure as compared to the microwave wavelength (Touzi et al. 2007; Ramsey et al. 2011). As such they are a great adjunct to visible and infrared sensors in teasing out the vagaries of the wetland variability.

The integration of other sensors such as LIDAR (http://lidar.cr.usgs.gov/knowledge.php; accessed July 21, 2012) has been particularly useful by adding a topographic variable as well as a canopy height variable (Ramsey and Jensen 1995; Ramsey et al. 1998; Ramsey et al. 2004; Yang et al. 2009; Yang and Artigas 2010; Ramsey et al. 2011). These are in addition to the types of information that RADAR supplies. Most of these studies also remind the gentle reader of the great import of ancillary and field data collection (Wright and Gallant 2007).

A number of techniques and approaches have been presented. The goal here is to present methods and approaches to enhance remote sensor data to detect, navigate, and characterize wetlands. However, there are only so many cartographic maps available. To provide additional ingests to analyses necessitates development of information from other means and likely remote sensor sources. Remote sensor analyses can provide information for the GIS and modeling stages of the work (for example, Yi et al. 1994; Maidment and Djokic 2000; Lopez et al. 2003; Lyon 2003; Martz and Garbrecht 2003; Mehaffey et al. 2005). Remote sensor products can be assessed as to their accuracy and precision, and land cover types can be addressed through producers and users accuracy furthering their utility in case studies (Congalton and Green 2009; Burnicki 2011).

It remains to demonstrate the uses of remote sensor data and fieldwork in wetland landscape ecology applications (for example, Van Derventer 1992; Shuman and Ambrose 2003; Ward and Trimble 2004; Stevens and Jensen 2007). An effective way to do this is through case studies where the integration of tools for a result is demonstrated.

Case Study: Great Lakes Wetland Characterization and Indicator Development

This work and case study demonstrate the role of landscape data in spatial and temporal wetland characterization. By identifying strategies for the assessments of the extent, composition, and vigor of wetland complexes at a synoptic scale the value of landscape ecology can be evaluated. The landscape metrics used were quantifiable measurements of wetland attributes, based on data that were spatially explicit and geographically referenced. In the context of wetland assessment and management, there are several areas where landscape metrics are useful. It is important that landscape data be linked to field data for both validation and scaling. Integration of landscape-scale metrics with ground-level wetland functional assessments is critical to managing natural resources, such as the Great Lakes. Key elements to consider are:

- Wetland inventory—A comprehensive assessment of the location and extent of wetland area that exists over the landscape is typically expressed in a geospatial context as a census of a particular time period (e.g., one year). A comprehensive wetland inventory provides a sampling *frame* from which particular sites can be selected for ground-based assessment and monitoring. It also can provide estimates of various types of wetlands and how they may change over time. A wetland inventory that consists of maps and statistics can also provide a reference to assist local, state, tribal, and federal agencies in evaluating projects for which they have permitting and oversight responsibilities. A census of wetland identification from ground-based surveys over the entire Great Lakes basin zone is impractical and has never been attempted; instead aerial photography and satellite sensors have been used to generate wetland inventories in the past.

- Wetland condition—Many measurements aimed at assessing wetlands conditions are obtained from ground-based sampling (e.g. water quality, biotic assemblages) and are described in other chapters. However, some aspects of wetland conditions can only be assessed using remote sensing. For example, changes in spectral reflectivity that indicate vegetative stress can only be assessed synoptically using remote sensing. In addition, the prevalence of some invasive or opportunistic plant taxa can best be assessed comprehensively using remote sensing, which also provides ideal tools for looking at spread over time. Recurring remote sensing assessments can also provide a means to monitor wetland loss, hydrologic alterations, changes to physical habitat condition, and other types of wetland change.

- Wetland setting—Natural aspects of the landscape in which wetlands exist (e.g., hydrology, climate, surface geology) significantly affect their physical and biotic characteristics. Responses to anthropogenic activities in the landscape (e.g., agriculture and development) are a major cause of wetland loss and degradation. Anthropogenic stressors are frequently physically removed from wetlands, with influences exerted over relatively large areas. Information on climate (e.g., temperature, precipitation), surface geology (e.g., soils), and watershed characteristics (e.g., flow direction, volume, and duration) are often available as geospatial data themes. Landscape data on anthropogenic stressors are widely available and can be used to assess both the intensity and types of impacts and spatial variability. Adjacent land cover may also be relevant to wetland conditions, as natural lands surrounding wetlands can buffer their vulnerability to anthropogenic impacts.

- Wetland and landscape spatial configuration—The spatial configuration of wetlands (i.e., size, shape, and interspersion within the larger landscape) can be important in regulating their function and conditions. Some questions require the consideration of wetlands as an interconnected suite rather than in isolation. For example, wetlands collectively support biodiversity over large areas, a function that is dependent on wetlands' connectivity and diversity in size, type, and composition. Landscape metrics that lend themselves to the use of remote sensing include those describing connectivity or spacing among wetlands, fragmentation, or heterogeneity of land cover categories, and patch size, patch shape, and patch interspersion.

Data needed to compute these types of landscape metrics are either obtained directly from remote sensing (e.g., land use/land cover maps based on interpretation of aerial images or categorization of satellite imagery) or assembled via spatial interpolation of ground-based surveys (e.g., census maps, soil surveys, bathymetry). For individual wetlands or over a small area, these data might be collected by a specific group using limited resources and techniques. On a broader scale, the cost, effort, and technical expertise needed to collect, analyze, and maintain such spatial data sets are typically beyond the resources of some single organizations. Consistent, repeatable, and broadly applicable data sets are needed to facilitate assessment of wetland conditions across the Great Lakes coastal zone. Even if local-scale assessments could be universally conducted across the region, a common protocol for landscape data would be needed to ensure that such data could be merged to create regional inventories and assess status and trends. Reliance on geospatial data and airborne/satellite imagery obtained and processed by collaborating government entities or commercial enterprises is often the only means for assessing landscape changes across the region.

Landscape Ecology Approach to Wetland Characterization

What do we mean by *landscape ecology,* and what are the fundamental elements of this scientific discipline? The interdisciplinary science of landscape ecology examines the distribution (i.e., patterns) of ecological communities or ecosystems, the ecological processes that affect those patterns, and changes in both the patterns and processes over space and time (U.S. Environmental Protection Agency 2001a). Sometimes the broader context of land and conditions surrounding the ecological communities or ecosystems of interest is referred to as *the landscape,* which serves as a conceptual unit for the study of spatial patterns in the physical environment and the influence of these patterns on important environmental resources.

Although basic ecological theory and concepts are underlying landscape ecology, it is different from some fundamental elements of traditional ecology because it takes into account the spatial arrangements of the components or elements that make up the environment. Landscape ecology analyses also account for the fact that some relationships between ecological patterns and processes can change, depending solely upon the scale at which the observations occur. The discipline of landscape ecology also includes the analyses of both humans and their activities as integral parts of the environment.

Thus a landscape is not solely defined by its size but by an interacting mosaic of elements (e.g., ecosystem types), which is relevant to some phenomenon or ecosystems of interest, such as wetlands and their ecological functions and services. Landscape ecology provides the ideal theoretical framework for analyzing spatial patterns relative to ecological condition and risk when it is desirable to assess a vast area, such as the Great Lakes Basin.

The landscape ecology approach presented in this case study is one of the techniques available for developing indicators of wetland conditions, which can be used for addressing real-world challenges of protecting and restoring such areas in a multitude of other regions, while assisting in the formulation of solutions that are beneficial to the public.

Applications

For a variety of reasons, some regulatory agencies and research communities have found it a challenge to complete basinwide assessments of Great Lakes wetland conditions. Some of the difficulty is related to the paucity of useful information, particularly in the critically important coastal areas, resulting in data for less than half of the ecological measures that have been identified by the 1998 State of the Lakes Ecosystem Conference (SOLEC) as important to monitor coastal wetland health. This fundamental shortage of comprehensive information about coastal wetlands is at the heart of the reason why there is no comprehensive long-term strategy for assessing the condition of Great Lakes coastal wetlands, an assessment of environmental impacts from development on coastal wetlands, or an assessment of the cumulative net (historical or projected) change of coastal wetlands. The U.S. Environmental Protection Agency (USEPA) is using the "landscape ecology approach" (Brown et al. 2004) to investigate and potentially resolve these outstanding questions about the status of coastal wetland resources in the Great Lakes, specifically by testing selected landscape metrics as potential indicators of ecological conditions in coastal wetlands.

A variety of regional, national, and international natural resource managers and decision makers, seeking to better understand the potential ecological impacts from loss of upland and wetland ecosystems in the Laurentian Great Lakes, at a broad scale, have adopted much of the broad-scale *landscape approach* described in this case study, led by the USEPA's Great Lakes National Program Office and the Great Lakes Commission, and other partners. The landscape approach has been actively used in the planning and implementation of wetland restoration research and projects, funded most recently through the Great Lakes Restoration Initiative (GLRI). Additional information about the ongoing Great Lakes Restoration Initiative is available at www.glri.us/ (accessed July 21, 2012).

A key aspect of the landscape approach in the GLRI and other restoration efforts in the region is the novel use of satellite and airborne remote sensing, the use of available and the development of new geospatial data products, and improved accuracy assessment procedures, linked by the theoretical construct of the "landscape ecology approach" (Brown et al. 2004). There are numerous other applications of this approach, which have allowed for the characterization of wetlands and their surrounding landscape conditions and processes around the world.

Researchers have previously used the landscape ecology approach to conduct simple regional assessments of environmental conditions, using some of the assessment results to further their goals of determining the interaction between landscape patterns and the flow of water, energy, nutrients, and

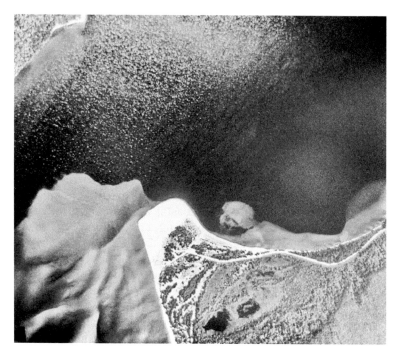

FIGURE 5.7
Connectivity among elements in the landscape can be demonstrated easily and quickly with remote sensing products of all types. The use of tone, pattern, and texture along with different scales and sensors can yield great information if you train yourself to interpret the characteristics along with field knowledge to converge on a solution or answer. This last image in the series also shows how deep waters reflect little light as compared to beach sands at the tip of the point. Sediment entrained in the surface water to the left of the point can provide a reflectance and acts as a water colorant to show turbulent water characteristics. To the right of the point, the same sediment material can reflect from the bottom to the sky in shallow waters showing bottom characteristics.

biota in the environment (e.g., Fauth et al. 2000; Turner et al. 2001; USEPA 2001b). Data about the size, shape, and connectivity of ecosystems or human-built areas have also been used to provide measurements that may be useful for indicating (i.e., by geospatial statistical inference) the condition of other things on the ground, for example, the condition of coastal wetlands in a particular region of the Great Lakes (Figure 5.7). Good indicators can reveal dominant ecological changes with the most efficient use of resources, but cannot be used to determine the ecological condition at very fine scales, for example, a specific coastal wetland reserve. Using geospatial statistical models and incorporating our existing knowledge (from empirical studies in coastal wetlands), measurements from the broad scale can be related to conditions in specific ecological resources, and used to verify that the landscape scale measurements are indeed an *indicator* of the ecological conditions on the ground. Thus, landscape metrics of ecological condition can provide a *basis* for assessments of ecological condition and can be substantiated using scientific methodologies. Caution should be exercised when contemplating the use of landscape ecological results to make decisions at scales other than that of the original input data.

Metrics and Indicators

Standard measurements of ecological resources provide ecological *metrics*. When measured at a relatively broad (i.e., "landscape") scale (Forman 1995), ecological metrics (such as the percent cover of cattail in a particular coastal wetland location) can be described as a *landscape metric*, that is, a measurement that describes the condition of an ecosystem's critical components (O'Neill et al. 1992). Calculation of landscape metrics (typically derived from information on spatial form or structural relationships) requires the use of spatial data, often displayed as a thematic map and contained within a GIS. There are many formats of thematic maps and several possible GIS platforms to select from. The primary uses of landscape metrics are the characterization of historical and current ecological condition, based on land cover information, with the possible extrapolation of current and past information to make predictions about the future of environmental conditions. The combination and analyses of past, present, and future ecological conditions are referred to as ecological (or land cover) change analyses.

Indicators can be thought of as pieces of evidence, or clues, that give us information about the condition of some environmental feature of interest (Great Lakes National Program Office 1999). Indicators have significance far beyond the actual values of the attribute measured. An indicator is a value calculated by statistically combining and summarizing relevant data. For example, doctors use human temperature and weight to gauge human condition, and economists use interest rates and unemployment to assess the status of economies. Economists make seasonal adjustments for these indicators with a model, and most look at several indicators together instead of just one at a time. Similarly, environmental indicators provide pieces of information that may tell us something about the true condition of our surroundings. An *ecological indicator* is defined as a sample measurement, typically obtained by collecting samples in the field of an ecological resource (Bromberg 1990; Hunsaker and Carpenter 1990). For example, collecting plant material in a coastal wetland for further measurements in a laboratory spectrometer may provide information about the amount of trace metals in the soil of the wetland, indicated by the concentration of those trace metals in the leaf of a wetland plant. The State of the Great Lakes Ecosystem Conference (2000) defines an indicator as: "A parameter or value that reflects the condition of an environmental (or human health) component, usually with a significance that extends beyond the measurement or value itself. Indicators provide the means to assess progress toward an objective."

Landscape metrics can therefore be used to characterize the environment at a broad scale, and they can be used to develop verified *landscape indicators*, including indicators of habitat quality, ecosystem function, and the flow of energy and materials within a landscape. Empirical ecological studies in coastal wetlands and other wetland ecosystems suggest that fundamental patch measurements (such as the size of wetlands) or processes (such as net primary productivity) may be suitable as landscape indicators of ecological condition in Great Lakes coastal wetlands.

It is important to remember at which scale a metric (ecological metric or landscape metric) is being applied so that the results of such analyses can be viewed in the context of actual conditions on the ground. Many land cover gradients are subtle, but the data used for the metric may not be appropriate for capturing such subtleties of the true gradient on the ground. For example, even though plants may be good indicators of soil trace metal concentrations in wetlands, field collection of 20 plant samples throughout a coastal wetland (analyzed in the laboratory to determine the concentration of trace metals in the leaves of each) may be inadequate to determine the concentration gradient(s) of trace metals across an entire wetland. This is similar to the problem that occurs at broader landscape scales with GIS data. If land cover data is provided at a 1 km pixel size, that resolution of GIS data may be too coarse to measure the true gradients on the ground (e.g., small wetlands may be missed). Thus two important guidelines for effectively using landscape ecology metrics and indicators at relatively broad scales are (1) select the most appropriate data for addressing the ecological process or *endpoint* of interest, and (2) select the geospatial model(s) that is (are) most appropriate for detecting or describing spatial or temporal change in the landscape. The selected landscape ecology endpoints and models can also be adjusted or modified to help interpret the measurements, and to better understand overall ecological conditions as improved data and understanding of ecological processes emerge.

Over time, landscape metric and indicator values can provide information on the trends in the condition of the ecosystem components, and provide key decision points. The information about trends helps to determine:

1. If it is necessary to intervene?
2. If so, which intervention will yield the best results?
3. How successful interventions have been?

Scale and Gradients

The term *scale* is generally defined by the *extent* of information and the *grain* of information. The extent of information is the spatial domain or the size of the area studied for which data are available (McGarigal 2002). The grain of information refers to the minimum resolution or size of the observation units, often identified as patches or digital picture elements (pixels). The pattern detected in any portion of the land is a function of scale. Landscape ecologists often consider the scale of the information they will use in their analyses and the gradients of land cover data or other biophysical data. In order to understand risks to ecological resources and humans, it is important to analyze the spatial patterns of environmental conditions on a variety of scales, for example, ranging from a single plot in a wetland to a large region, such as the coast of Lake Michigan. Scientists may select metrics and indicators that reflect environmental conditions on a variety of scales in both space and time. In this case study, *fine scale* refers to minute resolution, such as might be observed in a single plot at a particular wetland, and a *landscape scale* or *broad scale* refers to coarse resolution, such as images acquired by a satellite that might produce individual pixels that are 30 m on a side. A landscape ecological investigation requires a definition of the scale of the input data (e.g., 30 m pixel size for land cover), and requires the user to understand what scale is appropriate for their particular application (e.g., animal species requirements). It is an important responsibility of the user to exercise caution when attempting to make decisions at or among different scales of landscape ecological outputs.

For example, wetlands in the United States that are smaller than 900 m^2, or have a dimension less than 30 m, (i.e., the minimum pixel resolution of the current land cover GIS data) are too small to have been detected in the land cover classification process, and even slightly larger wetlands may be missed in the classification process because of factors related to the physics of the satellite sensor system used in the production of land cover data. Therefore broad-scale monitoring of such small wetland areas may be difficult by directly observed landscape metrics (Lopez et al. 2003).

It is important to select gradients (i.e., changes over space or time) of condition(s) that offer sufficient variability, and a sufficient number of field-sampling sites to compare among reporting units (Green 1979; Karr and Chu 1997; Lopez et al. 2002), in the event that ecological metrics are to be used to develop landscape indicators. Landscape (e.g., land cover) gradients may be useful for the development of landscape indicators because the statistical relationships between landscape metrics and ecological metrics can give clues about how two (or more) elements of the landscape may interact, such as the relationship between agriculture in a watershed and the concentration of phosphorus in wetlands and associated waterways. In addition, the use of previously observed *in situ* correlations between biophysical measurements may help to guide the analyses of relevant parameters that may be good indicators of ecological vulnerability at moderate to coarse scales.

Prior to the advent of GIS, it was prohibitively expensive and time consuming to calculate metrics of landscape composition and pattern at multiple (spatial or temporal) scales throughout a vast area of the landscape. Without a full understanding of the spatial and temporal patterns of landscape composition and pattern, the condition of coastal wetlands and the vulnerability of these resources to loss and degradation are limited. Landscape metrics can be correlated with ecological metrics collected in the field at a fine scale and, using statistical inference, these correlations can be used to determine the association between the broad-scale data (the landscape metric) and the fine-scale condition (the ecological metric). A determination of correlations between the broad scale (e.g., Riitters et al. 1995), moderate scale (e.g., Van der Valk and Davis 1980; Roth et al. 1996; Nagasaki and Nakamura 1999; Fauth et al. 2000; Lopez et al. 2002; Lopez and Fennessy 2002), and fine scale (e.g., Peterjohn and Corel 1984; Murkin and Kale 1986; Ehrenfeld and Schneider 1991; Willis and Mitsch 1995; McIntyre and Wiens 1999a; Luoto 2000) has not been completely explored. The current list of potential and operational indicators of condition for coastal wetlands (at several scales) of the Great Lakes Basin can be found in proceedings of the State of the Lakes Ecosystem Conference (SOLEC). Archived information about SOLEC is available at www.epa.gov/solec/ (accessed July 21, 2012).

Landscape Modeling

Because the landscape of the Great Lakes Basin is very complex, an initial focus on the relevant biophysical characteristics (i.e., excluding fewer relevant biophysical characteristics) is an important first step toward developing landscape indicators. GIS is a key tool that can be used to focus on relevant features of the landscape. For example, a GIS-derived landscape metric, such as percentage of cropland area among watersheds, can be correlated with water quality parameters at a location that is known to be the outlet of a watershed, and a geostatistical model can then be developed. The relationships might be analyzed as a causal (predictive) relationship, perhaps using a regression model with watershed condition as the independent variable(s) and water quality parameter(s) as the dependent variable(s). The causal relationships of these variables might be based on a priori knowledge acquired as a result of previously published in situ studies of similar variables and ecological theory as a whole. Broad-scale models founded on the ecological principals of in situ studies may be limited by a lack of detailed information about small areas, but can serve as a preliminary tool to assess large areas that would otherwise be impractical to assess in the field, or where a full coverage of detailed GIS data are absent. A specific and contemporary example of how to use remote sensing, GIS, and field-based techniques is demonstrated in other chapters, which may be used to determine the potential causal relationships between landscape disturbance, as described with broad-scale landscape metrics (the independent variables in this chapter). Many stressor variables may be tested in this manner (e.g., water quality measurements, habitat characteristics, or wetland functional characteristics), depending on the objectives of the user and the ecological endpoint of interest.

Selection and Use of Metrics and Indicators

Landscape metrics and landscape indicators (derived from the metric) may be used to assess progress toward one or more objectives (SOLEC 2000). Thus the selection and use of metrics and indicators should be guided by the purpose for which the information will be used, whether research oriented or policy oriented. Depending upon the use, the relative importance of quality, cost, and completeness of the coverage of the metric or indicator may differ for the user.

There is crossover between the goals a pure research landscape ecology and the policy uses of the results. The crossover between landscape ecology science and policy is a result of the common *primary* goals of each perspective, which are essentially focused on identifying key indicators of ecological condition that can serve as sentinels of important change. Despite this common primary goal, the differences between the scientific and policy uses of landscape ecology can be profound and is generally a result of differing secondary goals. The common primary goals of the research and policy communities are summarized next (Note: All topics are being addressed to some extent by both groups, although not always as a primary goal):

1. Assess changes in the condition of the ecosystems and the progress toward achieving management goals for its sustainable well-being.
2. Improve understanding of how human actions affect the ecosystems and determine the types of programs, policies, or regulations needed to address the environmental impacts.
3. Gain a clearer understanding of existing and emerging environmental problems and their solutions.
4. Provide information that assists the public and stakeholders in participating in informed decision making.
5. Provide information that will help managers better assess the success of current programs and provide a rationale for future ones.
6. Provide information that will help set priorities for research, data collection, monitoring, and cleanup programs.

Applications for Policy Goals and Regulatory Support

As environmental regulations were initially being developed in the United States, there was a focus on the established measures of environmental quality, such as those for drinking water and air quality. These measures reflected a traditional view of the environment and the potential for multiple factors that may contribute to environmental degradation. Research that was supported by regulatory agencies addressed the need to make policy recommendations to decision makers but did not fully address the scientific (i.e., ecological research) community's goal of increasing our understanding of the interrelationships between abiotic and biotic parameters (Zandbergen and Petersen 1995).

Thus landscape indicators were initially developed as lists of physical and chemical measures to monitor improvements in water quality. Biological responses resulting from changes were not considered. Requirements for environmental impact statements led to development of procedures to evaluate habitat as the basis for environmental assessment. As government policies endeavored to protect both human health and the environment from the byproducts of an industrial economy, scientific research required a different approach to support these policies. Awareness of the scope of environmental problems increased and toxic substances became a concern. A variety of tissue, cellular, and subcellular indicators was developed as diagnostic screening tools or biological markers to evaluate the physiological condition of an organism and to detect exposure to contaminants. The need to develop management strategies able to address interactions within ecosystems and the impacts of human activities upon those natural systems became another stimulus for the development of indicators.

The development and use of indicators that meet all of these needs is a learning process for both the scientists who develop them and for the policy makers who use them. Scientific knowledge itself is the outcome of a consensus-building process among scientists from different disciplines who require easily interpretable descriptions of ecological condition (Zandbergen and Petersen 1995). Developing landscape indicators involves the collection and management of supporting data, the identification and use of selection criteria, the evaluation of indicators for their efficacy, and accounting for the influence of scale on the final product. Landscape indicators are an important input to a decision support system, which can be utilized by policy makers and environmental professionals who require the most up-to-date and accurate information for determining effective strategies for ecological monitoring, assessment, restoration, characterization, risk assessment, and management.

Applications for Research

The current ecosystem concept and approach to studying ecological interrelationships was conceived as a multidisciplinary, problem-solving concept with the goals of restoring, rehabilitating, enhancing, and maintaining the integrity of particular ecosystems. The answers to scientific questions posed within ecosystems have created a new list of scientific questions that focus on how these ecosystems interact with the surrounding biophysical environment and thus have spurred a new area of investigation into the pattern of land cover and the implications of that land cover pattern on the ecosystems that are *embedded* within the land cover (e.g., a coastal wetland that is embedded in a larger landscape of agricultural crop land). None of these relationships have been fully field tested across a vast area, yet many landscape indicators have been conceptually proposed, that is, developed from theoretical ecology (USEPA 2001a). Very few results are available that show comparisons of landscape metrics or metric performance at different scales (Cushman and McGarigal 2004), but some of these relationships have been preliminarily analyzed, and several new patterns have recently been explored throughout the entire Great Lakes Basin for coastal wetlands.

The following indicators are among those evaluated by aggregate constituencies in the Great Lakes Region and include biophysical measurements that directly relate to ecological endpoints within wetlands of the Great Lakes as well as numerous other wetland regions of the world.

1. Amphibian community condition
2. Areal extent of wetlands by type
3. Bird community condition
4. Contaminant accumulation
5. Extent of upstream channelization
6. Fish community condition
7. Gain in restored wetland area by type
8. Habitat adjacent to wetlands
9. Human impact measures
10. Invertebrate community condition
11. Land-use classes adjacent to wetlands
12. Land-use classes in watersheds
13. Phosphorus and nitrogen levels
14. Plant community condition
15. Proximity to navigable channels
16. Proximity to recreational boating activity
17. Sediment flow and availability
18. Water level

Several basic criteria are generally accepted as the important ones to apply when selecting landscape indicators that are applicable to Great Lakes and associated wetlands, as well as other watersheds and associated wetlands of the world.

1. Cost and level of effort to implement basinwide
2. Measurability with existing technologies, programs, and data
3. Basinwide applicability or sampling by wetland type
4. Availability of complementary existing research or data
5. Indicator sensitivity to wetland condition changes
6. Ability to set endpoint or attainment levels

TABLE 5.1

Summary of Flora and Fauna Indicators and Their Methods

Indicator (SOLEC ID)	Measurement Description	Method Summary
Invertebrate community health (4501)	Diversity indices, adult caddisfly presence/absence and diversity.	Sweep nets, activity traps, backlighting, Hester-Dendy samplers. Need standardized processing. Need standardized habitat sampling. Repeat visits.
Fish community health and DELTs (4502, 4503)	Several diversity and abundance (fish per meter) measures, incidence rate of DELTs (deformities, eroded fins, lesions, and tumors).	Electroshocking along transects, fyke nets.
Amphibian diversity (4504)	Many possible population, diversity, and abundance measures. Compare with extensive measures. Species presence, abundance, and diversity.	From most intensive to most extensive–complete counts, capture-recapture, larvae sampling, drift fences or pitfall traps, funnel trapping, visual encounter surveys, Marsh Monitoring Program, and audio surveys.
Bird diversity and abundance (4507)	Intensive–many population, diversity, and abundance measures. Compare with extensive measures–species presence, abundance, and diversity.	Intensive–territory mapping, strip censuses, nest counts, site inventories. Extensive–MMP survey.
Plant community health (4513)	From air photos: % dominant vegetation types, % invasive types; from floristic survey: % wetland obligate species, % native taxa, floristic indexes; from quantitative sampling: % cover of invasives in dominant emergent, % floating/submersed cover of turbidity tolerant taxa, rate of change in invasive taxa.	Air photo compilation and interpretation, floristic survey, and quantitative sampling.
Contaminants (4506)	Contaminant levels or physical anomalies. Further work is needed to develop this indicator.	External survey of bullheads, DELTs, or other methods that provide useful biological contamination metrics.

Acquisition of the predominant information about a number of indicators relating to physical characteristics of Great Lakes wetlands and their surrounding environment can provide for integrated flora and fauna measurements. Indicator summaries in Table 5.1 and Table 5.2 can also be useful in the initial conceptualization stages of developing landscape ecological

TABLE 5.2

Summary Physical Characteristics and Methods for Obtaining These Measurements in Coastal Wetlands

Indicator (SOLEC ID)	Measurement Description	Method Summary
Water levels (4861)	Lake levels, wetland water levels, inflows/outflows.	Data obtained from lake gauges.
Sediment flow (4516)	Suspended sediment unit area yield (tons/km² of upstream watershed).	Metric should be estimated from gauging stations upstream of wetland. Sediment core or turbidity measures.
Sediment available for coastal nourishment (8142)	Sediment budget, net accumulation/loss.	Metrics measured from stream flow and sediment gauging stations at mouths of major tributaries. Alternatives– geomorphic surveys of barrier bars/islands, air photo interpretation.
Storms and ice	Possible metrics include wetland form factor, succession lag times, storm erosion of shore buffers; ice cover duration, ice thickness, ice jams.	Methods vary by metric.
Phosphorus and total nitrates (4860)	Total phosphorus and nitrates concentrations from May to July for correlation with other metrics. Further work is needed to develop this indicator.	Metric calculated from concentration and flow measures from gauging stations.

indicators of wetlands and their conditions by ensuring that geospatial data will adequately address ecological endpoints.

Field methodologies that are necessary to validate landscape gradients or to test landscape indicator sensitivity are included for the flora and fauna in Table 5.1 and for the physical measurements in Table 5.2. Field methodologies may be modified to directly address the ecological endpoint of the specific sensitivity of the landscape metric or indicator as needed.

Great Lakes Wetland Applications

For the North American landscape ecology metrics can be mapped among customary eight-digit hydrologic unit codes (United States) and hydrologic subsubdivision (Canada). The 1 km wide strip of area on the perimeter of the Great Lakes, where most of the remnants of the larger historical coastal wetlands exist today (Figures 5.8 and 5.9), are particularly important to quantify. The 1 km coastal region around the Great Lakes encompasses many, but not all, of the larger coastal wetlands in the basin. Although the

FIGURE 5.8
An oblique aerial view of a coastal wetland complex. Coastal wetland complexes like this are important landscape elements and their ecological functions provide many human services such as water quality improvement and flood attenuation. Wetland complexes can be described in terms of its wetland, open water, upland, and other biophysical components. An area of the landscape, such as the area depicted in this image, can be described as patches of ecosystems, vegetation associations, and patch metrics such as size, topographic position, interspersion, orientation, and relative proximity to components in the landscape.

1 km coastal region may not entirely capture all of the coastal wetlands within the basin, it is a useful approach for inferring the potential for disturbance of some of the landscape metrics that describe land cover that can directly affect coastal wetlands (e.g., road density and agricultural land cover metrics).

The 1 km coastal region was also included to provide recommendations for SOLEC and USEPA's Great Lakes National Program Office, which have traditionally assessed the condition of the Great Lakes Basin within the 1 km coastal region. Other assessment units are possible, such as the 5 km, 10 km, or full hydrologic unit, however larger landscape assessment unit alternatives also have limitations for focusing directly upon coastal issues.

Each of the coastal regions where landscape metrics were reported is also divided among the different hydrologic units of the Great Lakes Basin so that the calculations can be easily viewed and compared among them. Because the relatively narrow coastal regions were indistinguishable at the broad scale and difficult to portray using a full-basin map (Figure 5.10), each of the metrics for coastal regions (where applicable) is reported by coloring the full hydrologic unit associated with that length of coastal area (i.e., to aid visual interpretation of the printed or displayed map).

FIGURE 5.9
The size, configuration, and connectivity of nonwetland areas within the landscape may provide important clues about the condition of wetlands in the nearby vicinity; for example, those land cover and land use types immediately adjacent to or nearby the coastal wetlands.

All maps of coastal and full hydrologic unit metrics are directly comparable, but the user must pay close attention to the legend to identify the scale of analysis to which the map refers. Most of the maps in this chapter describe landscape metrics within the full hydrologic unit and the 1 km coastal region to demonstrate the range of approaches for mapping the results of the case study.

Characterization of Coastal Wetland Ecological Vulnerability

Coastal wetlands are vulnerable to loss or degradation as a result of the interaction of naturally occurring conditions and human activities (Table 5.3). Wetlands that are degraded as a result of conditions may continue to function but at a reduced functional level. Not all wetlands remain after these functional changes occur with some coastal wetlands losing their ecological functions quickly and some ceasing to exist altogether (i.e., wetland loss). Wetlands may flourish in conditions that fluctuate in their conditional state; for example, some coastal wetlands depend on periodic changes between standing water and exposed soil, which tends to increase the diversity of plants, which in turn supports wetland-independent animal habitat. Thus periodic wetland disturbances may allow for the formation of relatively small, interconnected metapopulations, where gene flow between plant patches or wetlands maintains the genetic diversity that might otherwise

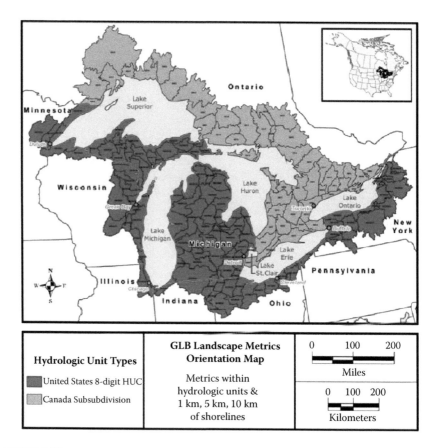

Hydrologic Unit Types	GLB Landscape Metrics Orientation Map	0 100 200
United States 8-digit HUC Canada Subsubdivision	Metrics within hydrologic units & 1 km, 5 km, 10 km of shorelines	Miles 0 100 200 Kilometers

FIGURE 5.10
Orientation map for Great Lakes case study area, utilizing a broad-scale approach for characterizing wetlands.

decline in relatively large inbred populations. When such populations become unable to bridge the gaps between populations, at the advanced stages of patch isolation, entire populations may become locally extinct (Opdam 1990). Water-level fluctuations also promote the interaction of aquatic and terrestrial ecosystems, and can result in higher quality habitat and increased productivity.

Environmental changes that can directly influence coastal wetland condition (such as dredging, filling, draining, and species invasion) which originate in the wetland itself are easier to pinpoint than indirect environmental changes. Direct environmental changes in coastal wetlands are often human induced, highly visible, and can result in rapid changes to wetlands. Indirect environmental changes are often less pronounced, potentially causing changes in wetland function and vegetation communities over a longer period of time. Indirect environmental changes are physically removed from

TABLE 5.3

Causes of Coastal Wetlands Loss or Degradation in the Great Lakes Basin

Adjacent urbanization
Change in magnitude and/or duration and/or frequency of water levels
Change in wetland vegetation, e.g., change in proportion of wetland open water and emergent vegetation
Chemical/oil spill
Dredging
Early ice breakup, early peaks in spring runoff, change in the timing of stream flow, and increased intensity of rainstorms
Habitat loss and fragmentation
Mechanical clearing of wetland vegetation
Overharvesting of resources
Reduced summer water levels
Removal of tree cover and shoreline vegetation
Runoff and pollutants from agricultural areas, sewage treatment outflows, stormwater outputs, urbanized areas, industrial outfalls, and other sources in watershed
Shoreline modification; wetland filling or drainage
Species invasion and spread (e.g., carp, zebra mussel, common reed, purple loosestrife)
Storms and seiches
Peak flows of runoff from paved urban areas may rapidly pulse through wetland and increase the amount of metals, oils, salts, or other contaminants into, or flowing out of, wetlands to open lake areas
Changed competitive or successional processes that may result in changed species diversity in fish, amphibian, bird, plant, or other community structure
Loss of optimum habitat for some species of fish, waterfowl, and other marsh birds
Death of wetland organisms
Deepening water and removal of sediments can result in loss of wetland habitat
Fewer viable breeding sites, especially for amphibians, migratory shorebirds, and waterfowl; northern migratory species (e.g., Canada geese) winter farther north; increased flooding frequency in coastal areas
Decrease in the available aquatic habitat for organisms, especially affecting species with limited dispersal capabilities (e.g., amphibians and mollusks)
Creation of impassable areas for some species, thus isolating populations and increasing likelihood of extirpation
Depletion of recreationally or commercially valuable species
Reduction in the total area of wetlands, resulting in poorer water quality and less habitat for wildlife
Increased runoff into wetland from adjacent land
Increased loading of nutrients, sediments, and toxic chemicals in downstream wetlands; reduced water clarity
Physical destruction or reduction in protection of coastal regions to erosion
Feeding, spawning, and nesting behavior of animals may interfere with plant photosynthesis/growth; nonnative animals may prey upon native animal species or outcompete them for food and habitat; plants may not provide suitable forage, nesting, reproduction
Damage to vegetation due to high winds and waves

a wetland, and thus it may be difficult to pinpoint the exact source of the environmental change. Indirect environmental changes include urban and agricultural runoff. Indirect environmental changes are relatively difficult to control, due to their diffuse and variable sources (Environment Canada 2002).

Human-induced environmental change factors described in this case study are based on previously observed positive correlations between ecosystem degradation and amount of land cover conversion during road construction, road maintenance, and other human activities (e.g., Connell and Slatyer 1977; Van der Valk 1981; Ehrenfeld 1983; Johnston 1989, 1994; Scott et al. 1993; Poiani and Dixon 1995; Jenning 1995; Wilcox 1995; Ogutu 1996; Stiling 1996; Heggem et al. 2000; Lopez et al. 2002; Lopez and Fennessy 2002), which enable the resource analyst to better understand the spatial configurations of wetlands within the larger context of the landscape.

The spatial configuration of coastal wetlands (i.e., size, shape, and interspersion within the larger landscape) is an important consideration, since larger wetlands may be relatively more likely to persist in the face of environmental changes. Wetlands of various sizes also attract different species, and a range of sizes may increase the diversity of habitat types across a broad area. For example, some birds (e.g., black tern, Forster's tern, and short-eared owl) may require a sufficiently large size before they will make use of it for nesting (Environment Canada 1998). Mitsch and Gosselink (2000) have described wetlands as spatially and temporally dynamic habitats, and thus the boundaries of coastal wetlands could be affected by the combined geological and hydrological processes associated with erosion and deposition, changing biological processes in the process. Wetland size and proximity metrics used here are based on previously observed trends regarding the effects of patch size, patch shape, and the interspersion of ecosystems within the broader landscape for specific taxa in many different regions (e.g., MacArthur and Wilson 1967; Simberloff and Wilson 1970; Diamond 1974; Forman et al. 1976; Pickett and Thompson 1978; Soule et al. 1979; Hermy and Stieperaere 1981; Van der Valk 1981; Simberloff and Abele 1982; McDonnell and Stiles 1983; Harris 1984; McDonnell 1984; Moller and Rordam 1985; Brown and Dinsmore 1986; Dzwonko and Loster 1988; Gutzwiller and Anderson 1992; Opdam et al. 1993; Hamazaki 1996; Kellman 1996; Bastin and Thomas 1999; McIntyre and Wiens 1999a, 1999b; Twedt and Loesch 1999; Lopez et al. 2002; Lopez and Fennessy 2002).

Extent of Coastal Wetlands

As discussed earlier, the extent of coastal wetlands in the Great Lakes is generally limited to a narrow area along the shoreline, however, some relatively larger coastal wetlands may extend further inland than 1 km, but this approach may include noncoastal areas too. The differences between coastal wetland extent among watersheds and coastal region areas can be used to interpret other metrics and to prioritize all of the Great Lakes coastal areas for more detailed analyses (Figure 5.11).

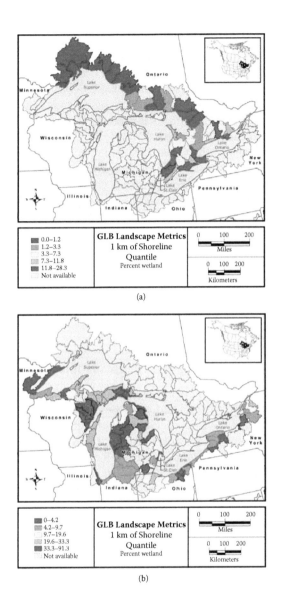

(a)

(b)

FIGURE 5.11
(See color insert.) Extent of coastal wetlands is mapped here as percent of wetland within the 1 km coastal region among coastal watersheds in (a) Canada and (b) the United States. Percent coastal wetland is calculated by dividing the number of wetland land cover cells in the coastal region of each watershed (i.e., the reporting unit) by the total number of land cover cells in the reporting unit minus those cells classified as water. This measurement has potential for measuring and comparing wetland contribution among watersheds and may be used to indicate potential for wetland removal or reduction in the amount of pollutants entering the Great Lakes. The relative extent of coastal wetlands may also be developed into a quantitative indicator of habitat for a wide variety of plant and animal species.

Prior to European settlement, the extent of wetlands in the Great Lakes Basin spanned large areas from the western edge of Lake Erie, across Ohio and Indiana, and covering the southern portion of the province of Ontario. It is estimated that two-thirds of Great Lakes coastal wetlands have been lost since European settlement. Many of these areas have been drained or reclaimed for land development, farmland, harbor facilities, and urban expansion (Environment Canada 2002). Other substantial wetland losses inland from the coast of the Great Lakes may contribute to the degradation of coastal wetlands as a result of suburbanization, dam construction, stream alteration, and the construction of flood control structures that alter the hydrology of contributing watersheds (Cox and Cintron 1997).

Between the 1780s and the 1980s, the largest reductions of coastal wetlands occurred in Ohio (USEPA 2001b). Urban development along the shores of the Great Lakes generally reflects the history of human decision-making processes that necessitated safe and efficient harbors for the distribution of natural resources, such as timber and mineral ores. As a result of these decisions, the areal extent of coastal wetlands has been dramatically reduced by the conversion to, and to some extent by the indirect effects of, urban and agricultural land use.

Coastal Interwetland Spacing and Landscape Integration

Interconnected wetland patches function as a network (e.g., within a watershed or migratory bird flyway) and have the cumulative functional capability of all the individual wetlands. A collection of wetlands in the landscape may be particularly important for providing a vital ecological unit for some animals, whereas other animals may require a mixture of wetland and upland areas for different portions of their life cycle or their daily activities (e.g., a species that reproduces in wetlands and forages in upland areas). The absence of such wetland complexes or integrated upland and wetland conditions may completely interrupt or degrade the reproduction rates, survival rates, and overall fitness of some plant and animal species.

Fragmentation of the landscape may result in the isolation of coastal wetlands, with the remnants of the formerly larger interconnected wetland complexes being replaced by less heterogeneous landscapes that are dominated by agricultural land, urban or rural human habitations, or industrial land. Such conversions of wetland to other land cover types may reduce the functional capability of coastal wetlands and may have also increased the likelihood that the remaining wetlands are further affected by the new land cover type (Tiner et al. 2002). Thus, as the general concept of ecosystem integrity describes, the capability of coastal wetlands to continue to function and provide ecological services to the residents of the Great Lakes (e.g., improving and maintaining clean water; providing critical habitat for plants and animals; and shoreline stabilization and protection) is dependent upon the effects of the surrounding landscape.

Wetland interconnectivity is one way of measuring the fragmentation of coastal wetlands in Great Lakes coastal regions (Figure 5.12). A standard and uniform method for measuring wetland interconnectivity in the coastal region (e.g., within 1 km of the shoreline) is to determine the probability a wetland area cell having a neighboring wetland, using a *moving window* over a GIS data set (i.e., a 9 pixel window × 9 pixel area as used in Figure 5.12). Thus the boundaries between all pixel pairs, where at least one pixel is categorized as wetland, were examined in the moving window. The interconnectivity metric is the number of boundaries where both pixels are wetland divided by the total number of wetland boundaries (regardless of neighbor land cover type). This metric gives a measure of how well the wetland is connected within the window sample area, with high values being better connected than low values.

The relative percentage of *perforated* wetland is another measurement of ecosystem fragmentation and is calculated here by using a moving 270 m^2 window (i.e., 9 pixel × 9 pixel) across the GIS land cover data set (Figure 5.13). When the percent wetland in the window is greater than 60% and greater than the window's mean wetland connectivity value, the wetland cell in the center of the window was categorized as perforated. The number of perforated wetland cells in the reporting unit was then divided by the reporting unit's total land area (i.e., the total number of cells in the reporting unit boundary minus those cells classified as water) to derive the percentage of perforated wetland. Perforated wetland generally consists of a patch of wetland with center upland area(s), such as would occur if small clearing(s) were made within a patch of wetland or if an area of wetland contained an interior upland region. Perforated wetlands may be fragmented in this fashion to such an extent that they do not provide suitable interior habitat for some wetland species. However, the interspersion of upland and wetland conditions in perforated wetlands may provide suitable habitat for some specialized plants and animals that require fluctuating wetland conditions and isolated upland areas. Thus high perforation values may be considered as detrimental for some ecological functions and species, and advantageous for others.

Fragmentation of coastal wetlands may lead to increased interwetland distances because of the increases in the incidence and extent of other land cover types developing in the intervening spaces (e.g., farm land or human habitations). Accordingly, mean distance to closest like-type wetland (Figure 5.14) is an important metric because it may indicate the likelihood of nearby similar wetland habitat (e.g., neighboring emergent–emergent wetlands for migratory bird resting and foraging or neighboring forest–forest wetlands for migratory song bird resting and foraging). The mean (for a reporting unit) minimum distance to closest wetland patch, for example, is the distance from each wetland patch to its nearest neighboring wetland patch, which should be measured from one patch edge to another patch edge, and may consist of multiple measures (e.g., mean of three nearest patches). This metric is useful in determining

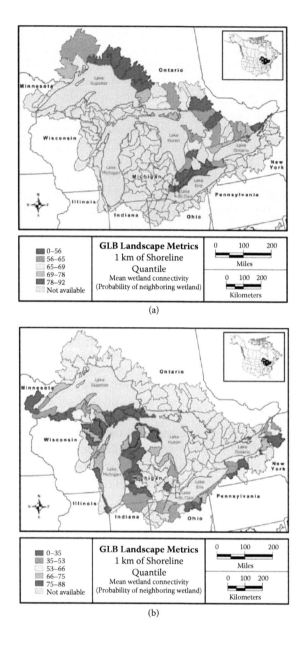

(a)

(b)

FIGURE 5.12

(See color insert.) Mean wetland connectivity in a 1 km coastal region of the Great Lakes Basin (probability of neighboring wetland), which is the mean (for a reporting unit) probability of a wetland cell having a neighboring wetland cell, calculated using a moving 270 m^2 window (9 pixels × 9 pixels) across the GIS land cover data set. Because these analyses use two differing land cover data sets, results for (a) Canada and (b) the United States may not be directly comparable.

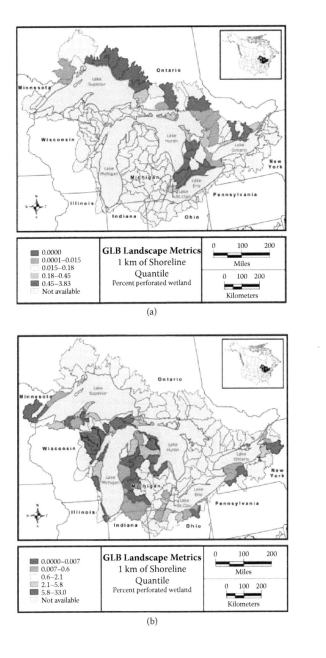

(a)

(b)

FIGURE 5.13

(See color insert.) Percentage of perforated wetland, in a 1 km coastal region of the Great Lakes Basin, is calculated using a moving 270 m² window (9 pixels × 9 pixels) across the land cover, and generally indicates if center upland area(s) are present in a wetland. Because these analyses use two differing land cover data sets, results for (a) Canada and (b) the United States may not be directly comparable.

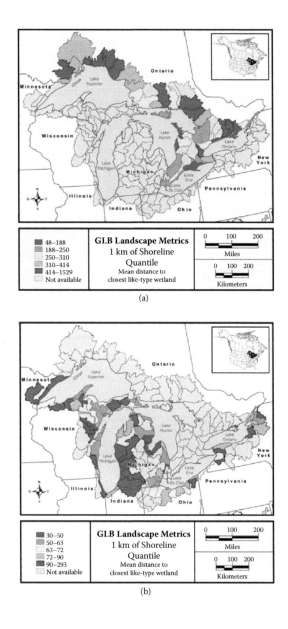

FIGURE 5.14
(See color insert.) Mean distance to closest like-type wetland, in a 1 km coastal region of the Great Lakes Basin, is the mean minimum distance to closest wetland patch, for the 1 km shore area, within each hydrologic unit. Distances were measured from edge to edge and are reported in meters. This metric is useful in determining relative wetland habitat suitability at scales that are ecologically meaningful for specific plant and animal taxa. Because these analyses use two differing land cover data sets, results for (a) Canada and (b) the United States may not be directly comparable.

relative wetland habitat suitability at scales that are ecologically meaning-ful for specific plant and animal taxa, and demonstrates the importance of establishing the ecological endpoint(s) of interest prior to full develop-ment of this indicator.

The Shannon–Wiener index and Simpson's index are two different ways of measuring the diversity and distribution of land cover types within a spe-cific area of the landscape. The Shannon–Wiener index of land cover type diversity (Figure 5.15) is calculated as

$$m\ H = -\sum Pi * \ln Pi, i = 1,$$

where Pi is the proportion of land cover type i.

Shannon–Wiener index values increase as the number of land cover types within the reporting unit increases, with higher value coastal areas having more diverse land cover (i.e., more diversity) than areas with lower values. Because higher Shannon–Wiener diversity in coastal areas does not always indicate greater opportunities for variety of species (i.e., land cover diver-sity includes agriculture and urban), Simpson's index (Figure 5.16) can be used to better describe the distribution of the land cover in a coastal region. Simpson's index is a quantitative measure of the evenness of the distribution of land cover classes and is most sensitive to the presence of common land cover types within a reporting unit. Simpson's index values range from 0 to 1, with 1 representing perfect evenness of all land cover types within a reporting unit. Simpson's index is calculated as

$$m\ C = 1 - \sum Pi^2, i = 1,$$

where Pi is the proportion of land cover type i.

Proximity of Land Cover and Land Use to Coastal Wetlands

The coastal region of the Great Lakes has been an attractive location for development during the history of settlement and expansion of societ-ies. The shorelines were a focus of human activities because they are near water, which provides unique transportation functions, resources for manu-facturing, recreational opportunities, residential uses, and drinking water resources. The transportation services in combination with the close prox-imity to productive farmland, raw materials, and an ever-growing inland infrastructure makes the coastal areas historically been an unparalleled area to economically exploit. Thus there may be conflicts between preserving the remaining coastal wetlands and maintaining and/or developing these areas for additional commercial and societal needs. Coastal wetland areas that are close to urbanization (Figure 5.17) or human population centers (Figure 5.18) may be sensitive natural areas and affected by human land use associated with urban and suburban activities.

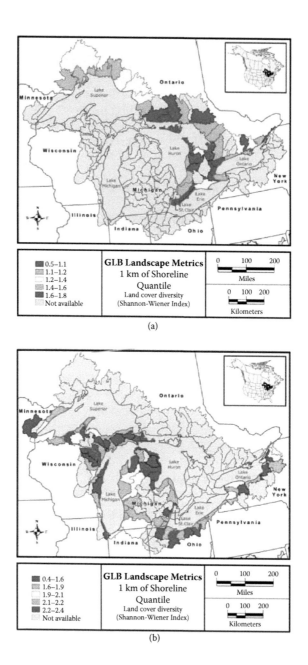

(a)

(b)

FIGURE 5.15

The Shannon–Wiener index, in a 1 km coastal region of the Great Lakes Basin, is one of several ways to measure the diversity of land cover types within a specific area of the landscape. The Shannon–Wiener index values increases as the number of land cover types within the reporting unit increases. Because these analyses use two differing land cover data sets, results for (a) Canada and (b) the United States may not be directly comparable.

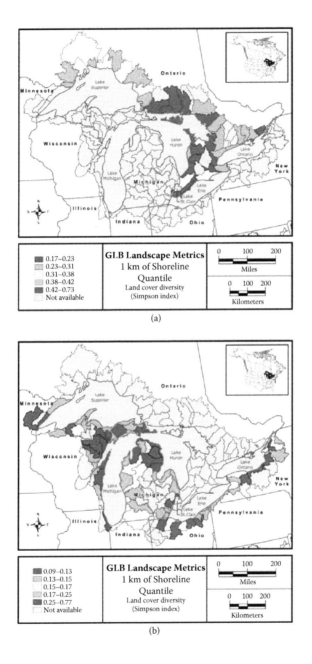

(a)

(b)

FIGURE 5.16

Simpson's index, in a 1 km coastal region of the Great Lakes Basin, is a measure of the diversity of land cover types within a specific area of the landscape. Simpson's index is a measure of the evenness of the distribution of land cover classes. Because these analyses use two differing land cover data sets, results for (a) Canada and (b) the United States may not be directly comparable.

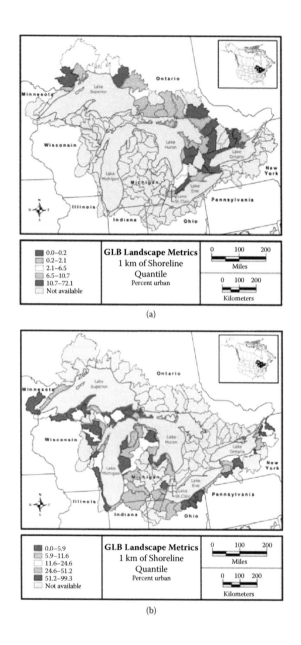

FIGURE 5.17
The percentage of urban land cover, in a 1 km coastal region of the Great Lakes Basin, is cal-culated by dividing the number of urban land cover cells in the reporting unit by the total number of land cover cells in the reporting unit minus those cells classified as water (i.e., total land area). High amounts of urban land indicate substantial modification of natural vegetation cover and may affect the condition of wildlife habitat, soil erosion, and water quality in coastal areas. Because these analyses use two differing land cover data sets, results for (a) Canada and (b) the United States may not be directly comparable.

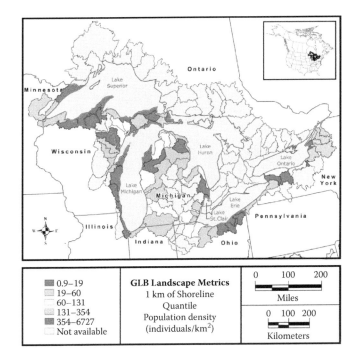

	GLB Landscape Metrics	0 100 200
■ 0.9–19	1 km of Shoreline	
□ 19–60	Quantile	Miles
□ 60–131	Population density	0 100 200
▨ 131–354	(individuals/km^2)	
■ 354–6727		
□ Not available		Kilometers

FIGURE 5.18

Human population density (individuals/km^2) approximated in the 1 km coastal region of the Great Lakes Basin (United States data only). Population density is calculated by summing the number of people living in the reporting unit and dividing by the reporting unit area. Where census units are not completely contained within the reporting unit, population is apportioned by area. High population densities are generally well correlated with high amounts of human land uses, especially urban and residential development. Large areas of development often involve substantial modification of natural vegetation cover that may have substantial effects on wildlife habitat, soil erosion, and water quality.

An example of the effects of coastal wetland disturbance is an increased expansion of invasive or opportunistic plants into landscape gaps (i.e., within and between wetlands), which may be the result of increased land cover fragmentation (Forman 1995). The patch dynamics (i.e., either increases or decreases in extent) of invasive and opportunistic plant species in disturbed Great Lakes coastal areas may be facilitated by the extent and intensity of wetland disturbance that results from human fragmentation of the landscape, resulting in hydrologic alteration (e.g., road construction). Because species-level assessments may not be possible using satellite or other coarse-scale remote sensing data (i.e., spatial or spectral resolution data), it may be necessary to also map invasive or opportunistic species using finer scale remote sensing data, such as was described in Chapter 4. The mapping approach described in this chapter provides powerful approaches for determining potential causal relationships between landscape disturbance determined

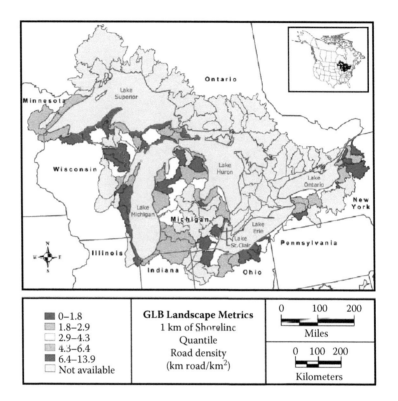

FIGURE 5.19
Road density (km road/km²) in a 1 km coastal region of the Great Lakes Basin (United States data only). The density of roads is calculated by summing the length of roads and dividing by the area of the reporting unit. Values are reported as length of all road types (i.e., freeways, highways, surface streets, rural routes, and other roadways) per reporting unit area. High total road densities are generally well correlated with high human population and urban development in the coastal region.

from landscape metrics and the receptor variables, such as those receptor variables associated with the extent and intensity of invasive organisms.

Data about the land cover and land use that is in the vicinity or directly adjacent to coastal wetlands may be important indicators of the level of disturbance within a wetland. For example, paved surfaces (e.g., roads; Figure 5.19) increase the impermeability (Figure 5.20) of land surfaces and may increase the amount of runoff to streams, lakes, and wetlands, and potentially increase the transport of road salts or other chemicals from paved surfaces (e.g., trace metals and hydrocarbons). Roads also fragment habitat and may act as barriers to animal movement (e.g., amphibians or large mammals).

Land use in a particular watershed may also have a significant influence on the flow of runoff and sediments toward coastal areas, and may be indicative of the amount of runoff that is intercepted by coastal (and other) wetlands. The capability of such wetlands to accumulate, transform, or store pollutants

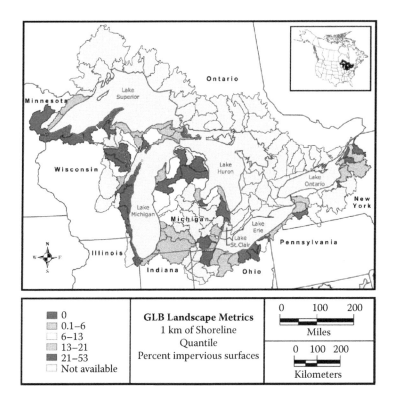

FIGURE 5.20
Percent impervious surfaces are mapped within a 1 km coastal region of the Great Lakes Basin (United States data only). The percent of total impervious area is calculated using road density as the independent variable in a linear regression model.

that are transported in the runoff from the inland areas of the watershed is an important mechanism for maintaining in order to improve the water quality of the Great Lakes Basin. Wetlands that are adjacent to other habitats and that provide connections between among other ecosystems in the watershed are also more likely to maintain their normal hydrologic regime, which may moderate the amount of water, sediment, and chemical constituents that are directly input into the open water areas of the Great Lakes Basin. Thus areas that are relatively more developed and intensively used for agriculture may have increased rates of runoff and sediment loading to the Great Lakes. However, if coastal (and other) wetlands, situated between upland urban or agricultural areas, are present in the landscape the runoff and sediment loading may be reduced. However, wetlands in close proximity to agricultural (Figure 5.21) or urban (Figure 5.22) land may be at greater risk of loss or degradation as a result of hypereutrophication or pollution. Wetlands that are adjacent to urban land cover may also provide poor animal habitat relative to wetlands adjacent to natural land cover, such as forests.

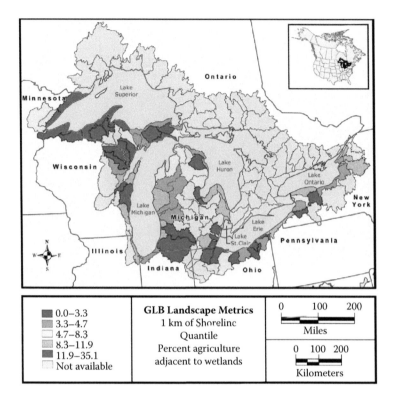

0.0–3.3	**GLB Landscape Metrics**	0 100 200
3.3–4.7	1 km of Shoreline	
4.7–8.3	Quantile	Miles
8.3–11.9	Percent agriculture	0 100 200
11.9–35.1	adjacent to wetlands	
Not available		Kilometers

FIGURE 5.21
(See color insert.) Percent agriculture adjacent to wetlands is mapped within a 1 km coastal region of the Great Lakes Basin (United States data only). The percentage of all agricultural land cover adjacent to wetlands is calculated by summing the total number of pasture and cropland land cover cells directly adjacent to wetland land cover cells in the reporting unit and dividing by wetland total area in the reporting unit.

Additional Applications

Several completed or ongoing landscape mapping programs in the Great Lakes serve as sources of landscape data for other geographies. These programs (Table 5.4) provide excellent guidance on the scale, scope, and procedural details of the key data necessary for a successful wetland characterization projects and programs.

Temporal Scale: Providing Consistent Measurements across the Great Lakes Basin

In general, major landscape metrics change slowly, compared with other types of response variables. The average return intervals for forest harvests, for example, are about 100 years, which translates to 1% of the landscape harvested on an annual basis. Rates of forest change due to natural disturbances

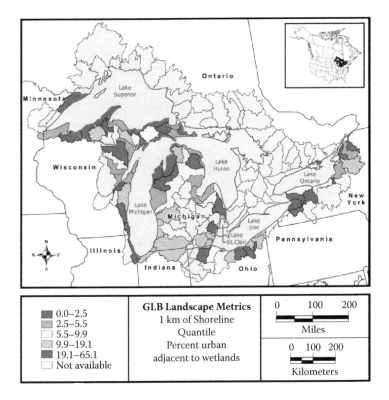

0.0–2.5	**GLB Landscape Metrics**	0 100 200
2.5–5.5	1 km of Shoreline	
5.5–9.9	Quantile	Miles
9.9–19.1	Percent urban	0 100 200
19.1–65.1	adjacent to wetlands	
Not available		Kilometers

FIGURE 5.22
(See color insert.) The percentage of urban land cover adjacent to wetlands is mapped within a 1 km coastal region of the Great Lakes Basin (United States data only). Percent urban is calculated by summing the total number of urban land cover cells directly adjacent to wetland land cover cells in the reporting unit and dividing by wetland total area in the reporting unit.

are even lower. Typical historic stand replacing fire return intervals range from 200 to 400 years (0.50%–0.25% annualized landscape change; White and Host 2008). Human-caused changes to the landscape, such as wetland loss or development, occur at greater rates but still involve relatively low percentages when landscape area is calculated on a basinwide or regional basis. As a result, regional landscapes can be effectively monitored over multiyear time scales; a five-year revisit interval is common for several agencies conducting integrated monitoring approaches. Unlike natural disturbances, however, human modification of the landscape tends to be spatially concentrated. The interface between urban–agriculture or urban–forest regions is one example of locations exhibiting rapid rates of land use change; the Great Lakes coasts and inland lakes are another. In areas of great human activity, both the spatial resolution of the source data and the temporal resolution of sampling frequency should be increased to more precisely track these changes; in these cases, a biennial revisit schedule may be appropriate.

TABLE 5.4

Available Landscape Data from Mapping and Characterization Programs

Historical Wetland Map	Resolution	Agency	Era	Extent	Base Data
U.S. National Wetlands Inventory (NWI), www.fws.gov/nwi/	.01–1 m	USFWS	1970s–present	U.S. nationwide	Aerial photos
National Land Cover Database (NLCD), usgs.gov/natllandcover.php	30 m	USEPA	2001	U.S. nationwide	Landsat
Coastal Change Assessment Program (C-CAP), www.csc.noaa.gov/crs/lca/ccap.html	30 m	NOAA	2001	U.S. coastal basins, lower 48	Landsat
Canadian Wetland Inventory, http://maps.ducks.ca/cwi/	30 m	Environment Canada	2000	Canada Nationwide	Landsat/Radars at
Minnesota CWAMMS, www.pca.state.mn.us/water/wetlands/cwamms.html	.01–1 m	Minnesota DNR & Pollution Control Agency	2006 present	Minnesota	Aerial photos and satellites
Wisconsin WWI, http://dnr.wi.gov/topic/wetlands/inventory.html	.01–1 m	Wisconsin DNR	1978–present	Wisconsin	Aerial photos
Wisconsin–WISCLAND, http://dnr.wi.gov/maps/gis/datalandcover.html	30 m	Wisconsin DNR	1992	Wisconsin	Landsat
Michigan Resource Information System–Current Use Inventory (MIRIS-CUI), www.mcgi.state.mi.us/mgdl/?rel = thext&action = thmname &cid = 5&cat = Land+Cover%2F Use+MIRIS+1978	.01–1 m	Michigan DNR	1978	Michigan	Aerial photos

Remote Sensing and Ancillary Data Sources

Remote sensors work in many regions of the electromagnetic spectrum from optical and ultraviolet to near-infrared to thermal and RADAR. Similarly, the resolutions vary widely among sensors from kilometers (AVHRR and MODIS) to a few meters (IKONOS). Some of the data sources are free to users (MODIS) or relatively inexpensive (Japanese Aerospace Exploration Agency [JERS] and PALSAR, US $25 and US $125 per scene), but generally speaking, the finer the resolution the higher the cost. The sensor choice depends on the study area, availability of ancillary data, cost, the resolution desired and what features need to be observed or monitored. To routinely monitor a large regional area such as the Great Lakes Basin, moderate resolution (e.g., 30 m grid cells or ¼ ha) would be the best choice. High-risk areas should be reviewed more closely with higher resolution imagery or air photos and field truth. Sometimes there are advantages to using coarser resolution data with a frequent (1–2 day) repeat, especially when looking for large-scale features (e.g., algal blooms can be seen in 1 km MODIS and AVHRR data) or more general regional changes due to climate (e.g., leaf area index and Fraction of Photosynthetically Active Radiation (FPAR) with MODIS products). Using repeat pass satellite imagery allows the advantage of multitemporal data analysis. In many cases, however, finer-scale but less frequently generated data are necessary. A monitoring plan using both high- and moderate-resolution sensors will provide the greatest amount of information.

Traditionally, optical and infrared data have been used for land cover mapping, including wetlands. However, wetlands are difficult to map and monitor using this type of data alone due to the high variability in wetland morphology and the inability of optical sensors to detect flooding beneath closed tree canopies. There are additional problems associated with cloud cover and obtaining data with optical systems during timely conditions. Some of the most promising new sensors for mapping and monitoring wetlands include those operating in the thermal and microwave spectra. Additionally, unlike optical, thermal, or infrared data, RADAR data can be collected during day or night and penetrate clouds so that timely data may be collected. Systems using LIDAR, synthetic aperture RADAR (SAR), and thermal infrared provide information complementary to optical sensors and will be invaluable in future mapping and monitoring programs.

Landsat Data and the Expected Future Data Gaps

The Landsat series of optical/infrared sensors has been widely used to study land cover processes. However, there is concern in the scientific community regarding the quality of the current information and availability of future Landsat data. The first Landsat sensor was launched in 1972, and the latest orbiting sensor, Landsat 7, was launched in 1999. A new Landsat Continuity Mission is scheduled for launch in 2013 (http://ldcm.gsfc.nasa.gov). The U.S.

government has also explored a program to provide alternative non-U.S. earth satellite imagery to current government and nongovernmental U.S.-based Landsat users. Alternate data sources for any potential gap include the Indian ResourceSat-1, which carries 23 m and 55 m ground resolution sensors; the Chinese/Brazilian CBERS-2, with several sensors on board ranging from 20 to 80 m resolution; France's SPOT sensors; the Japanese ASTER sensor, whose data are comparable to Landsat-TM but at 15 m resolution and with a smaller spatial extent; and other alternative remote sensing platforms. Additional information about platform and sensor options are well summarized by the U.S. National Aeronautics and Space Administration (http://rsd.gsfc.nasa.gov/rsd/RemoteSensing.html#list. This effort is meant to provide data sharing agreements and access to imagery that would otherwise be difficult for many users to obtain. The U.S. Department of Agriculture has already begun purchasing the Advanced Wide Field Sensor (AWiFS) imagery for its crop mapping activities over the United States.

Table 5.5 provides a (nonexhaustive) comparison of current resources available and provides an example of the necessary compilation that should occur prior to embarking on any broad-scale project, including information about sensor capability, cost, and other utilization notes that are important to the specific resource questions and decisions.

Improving Wetland Characterization through Ancillary Data and Processing

Several types of geographic ancillary data sets are useful for wetland mapping, including elevation data and soils maps, especially in combination with remote sensing data. Some new processing approaches (described next) can also be used to improve wetland mapping accuracy.

Elevation Data

There are typically one or more digital elevation model (DEM) data sets available for U.S. project areas, including the Great Lakes Basin. Canada DEMs are available from the Canadian government and other private sources. DEM utility depends on the resolution of the model, both in elevation and on the ground. Generally, the 30 m DEM available from the USGS is too coarse for the fine-scale microrelief that often results in wetland development. Interferometric SAR and LIDAR are two sources of remote sensing that can produce higher resolution DEMs.

Soils

The National Resources Conservation Service (NRCS) has created soils maps with classification of hydric soils that can be a useful ancillary data set in mapping wetlands. There are two U.S. soils maps sources: U.S. General Soil

TABLE 5.5

Current and Historical Sensor Data

Sensor	Frequency	Number Spectral Bands	Spatial Resolution	Size/Revisit Time	Operation Period	Cost	Source
AVHRR	MS	4–6	1.1 km	2400 × 6400 km single swath, other options	1978–present	Free (single scene); $190 stitched georegistered segments	http://noaasis.noaa.gov/NOAASIS/ml/avhrr.html
MODIS	MS	36; 0.4–14.5 μm	250–1000 m	2300 km; 1–2 day revisit	2000–present	Free	LPDAC; http://modis.gsfc.nasa.gov/
Landsat ETM+; Landsat TM; Landsat MSS	Pan/MS	8 bands 0.4–12.5 μm; 7 bands 0.4–12.5 μm 5 bands	15 m pan 30 m; 60 m thermal; 30 m 120 m thermal 60 m	18 km; 5–16 day revisit	1999–present; 1982–present; 1973–1983†	$425 TM; $700 ETM+	EROS http://landsat.gsfc.nasa.gov/
AWiFS	MS	4 bands 0.52–1.7 μm	56 m at nadir; 70 m at field edges	5 day revisit; 737 km swath	2003–present	$700/quad; 4 quads/scene	www.euromap.de/docs/doc_005.html; http://directory.eoportal.org/pres_IRSP6IndianRemoteSensingSatellite.html
LISS-III	MS	4 bands 0.52–1.7 μm	23.5 m	24 days; 140 km swath	2003–present		www.euromap.de/docs/doc_005.html; http://directory.eoportal.org/pres_IRSP6IndianRemoteSensingSatellite.html
ASTER	MS	15 bands 0.5–12 μm	15 m VNIR; 30 m SWIR; 90 m TIR	4–16 day revisit, 60 km	2000–present	Free–$170+	http://asterweb.jpl.nasa.gov/ Free (already existing Level 1B data over the U.S. and territories, available through the LPDAAC Data Pool)

(continued)

TABLE 5.5 (CONTINUED)

Current and Historical Sensor Data

Sensor	Frequency	Number Spectral Bands	Spatial Resolution	Size/Revisit Time	Operation Period	Cost	Source
SPOT	Pan/MS	4	10 m Pan/20 m MS	20 × 20 to 60 × 60 km	1986–present	$1,000–$14,000	www.spotimage.fr/html/_167_.php (already existing Level 1B 20 × 20 km, which is 1/8 scene, is 1,020 euro or about $1,400), no per km price found
IRS	Pan/MS	4–6	6 m Pan/23 m MS	148 km	1988–present	$375 per 10 × 10 km map sheet	Photosat, http://rst.gsfc.nasa.gov/Intro/Part2_23.html, http://ccrs.nrcan.gc.ca/resource/tutor/fundam/chapter2/12_e.php
Quickbird/IKONOS	Pan/MS	5 bands 0.45–0.9 μm	0.6 m pan; 2.4 m	3–7 day revisit; 16.5 km	2001–present	$8/km (IKONOS) $16/km (Quickbird)	www.digitalglobe.com/
Aerial Imagery, e.g., CAESAR™	MS	12 bands VIS	0.5–4 m	Weather and flight logistics		High	CAESAR was a NATO project that ended in 2005, to be replaced by MAJIIC; http://edcsns17.cr.usgs.gov/airborne/
Air photos	Pan/Color/Color-IR		0.1–1.0 m	Small spatial area/varies/as tasked	1909–present airplane	High	
OrbView 3™	Pan/MS	4 bands VIS/NIR	1 m pan 4 m	<3 days	2003	$10–$50/km	www.orbimage.com/corp/orbimage_system/ov3/; www.geoeye.com/whitepapers_pdfs/OV-3_Catalog.pdf
HyMAP Imaging Spectrometer™		128 VIS NIR SWIR	3.5–10 m	Weather and flight logistics		$6,000 per 2.3 × 20 km scene; proprietary data $12,000 per scene	www.hyvista.com

Sensor	Band	Wavelength/polarization	Resolution	Coverage/revisit	Date	Cost	Source
Fugro Earthdata LiDAR	Light (350–800 nm)		35 cm vertical, 3 m horizontal	As tasked			http://www.earthdata.com/servicessub cat.php?subcat = lidar
RADARSAT-1	C-band SAR	1, 5.7 cm C-HH Multiple modes and incidence angles	10 m (fine beam mode), 30 m, and100 m (wideswath mode)	45–500 km; approx. 6 day revisit; 50–500 km	1995–present	$0–$2,750	ASF, www.asf.alaska.edu/; MDA Corporation, http://gs.mdacorporation.com/
RADARSAT-2	C-band SAR	4 5.7 cm CHH, C-HV, C-VH, C-VV	3 to 100 m	500 km, daily to 3 days	2007 launch		http://gs.mdacorporation.com/; www.radarsat2.info/; www.space-gc.ca/asc/eng/satellites/ra darsat2/innovations.asp
ERS-1 and 2	C-band SAR	1, 5.7cm, C-VV 23 incidence angle	30 m	100 km; 35 day revisit	1991–present	$85–$450+	http://earth.esa.int/ers/Eurimage; ESA, http://eopi.esa.int/esa/esa?cmd = aodet ail&aoname = cat1 http://eods.nrcan.gc.ca/ers_e.php
Envisat	C-band SAR	2, 5.7 cm Any 2 : C-VV,CHH, C-VH, C-HV Multiple incidence angles	10, 30 m, and 100 m (wideswath mode)	100–400 km 35 day revisit	2002–present	$480 (archive); $720 (new)*; $125 (ESA CAT-1 data grant)	Eurimage* and ESA, http://earth.esa.int/ers, http://eopi.esa.int/esa/esa?cmd = aodet ail&aoname = cat1
JERS	L-band SAR	1, 23 cm; L-HH 37 incidence angle	30 m	70 km	1992–1998	$25	JAXA's CROSS, https://cross.restec.or.jp/cross/CfcLogin. do?locale = en
PALSAR	L-band SAR	4, 23 cm, L, HH, L-HV, LVH, L-VV	7–100 m	40–350 km	2006–present	$125	ASF, www.palsar.ersdac.or.jp/e/index.shtml
Airborne Radar GeoSAR	X-, P-band SAR	3 cm X-VV 86 cm P-full polygon	1.25–3 m (X) 1.25–5 m (P)	20 km	2002–present		http://southport.jpl.nasa.gov/html/projects/geosar/geosar.html www.earthdata.com/servicessubcat.ph p?subcat = ifsar

(continued)

TABLE 5.5 (CONTINUED)

Current and Historical Sensor Data

Sensor	Frequency	Number Spectral Bands	Spatial Resolution	Size/Revisit Time	Operation Period	Cost	Source
Airborne Radar (AIRSAR)	C-, L-, P band SAR; Also TOPSAR (DEM)	4, 5.7 cm C-band full polygon, 25 cm L-band full polygon, 68 cm P-band full polygon	2.5–12 m	As tasked	1988–2005 not in operation; JPL will fly if commissioned	Free–$750	http://airsar.jpl.nasa.gov/documents/faqs.htm#p4, http://airsar.jpl.nasa.gov/main.htm
Airborne Radar Fugro Earthdata GeoSAR	X-, P-band IFSAR	3 cm X-VV; 86 cm P-full polygon	1.25–3 m (X) 1.25–5m (P) 3–5 m DEM 36 cm P. LIDAR Night or day, through clouds and vegetation	12–14 km swaths with up to 1200 km flight lines; revisit as needed	2002–present	$30 to $170/sq km, no licensing, clients free to share and use as they see fit	http://southport.jpl.nasa.gov/html/projects/geosar/geosar.html, www.earthdata.com/servicessubcat.ph p?subcat = ifsar

Map (STATSGO; http://soils.usda.gov/survey/geography/statsgo/) available for every state but at coarser scale and Soil Survey Geographic Database (SSURGO; http://soils.usda.gov/survey/geography/ssurgo/) available for all states except Alaska.

Considering Innovative Processing Approaches

Object-based classification or categorization methods, such as those available in the eCognition image processing software immediately, may be useful for wetland mapping. This type of process involves two steps: (1) spatial objects are formed using a region-growing segmentation algorithm to merge pixels of homogeneous type; then (2) image classification techniques are applied using traditional statistical methods, a fuzzy logic rule base, or a combination of both methods. The segmentation phase provides additional attributes describing the spatial context and morphology of features that can be used to inform the classification beyond spectral values alone. Moreover, the segmentation phase can be reiterated at various scales to capture the range of features contained in the image. This also allows heterogeneous wetland types (e.g., wetlands containing some open water pixels mixed with denser canopy) to be grouped or not depending on the scale of the segmentation. The operator makes the decision. Grenier et al. (2007) have applied this processing approach to Landsat/Radarsat mapping in Quebec (Canada) and describe the process and results in detail.

Moving Forward with the Landscape Perspective

As noted earlier, implicit within the major challenges for resource decision making are questions related to the question: How do I put it all together? With the amount of available characterization methods, traditional and contemporary; relevant field-based data, which are designed with broad-scale approaches in mind; and the landscape perspectives described and demonstrated in this chapter, the answer to this important question becomes more apparent. In the next chapter, the integration of these elements is described and demonstrated so as to pull together the many pieces of the wetland landscape characterization puzzle.

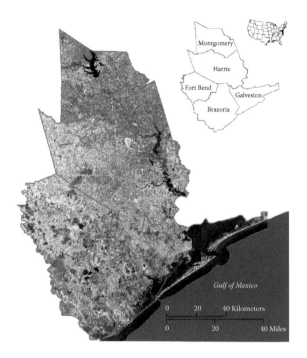

FIGURE 3.10
The study area is the combined area of Montgomery County, Harris County, Fort Bend County, Brazoria County, and Galveston County, Texas. The remote sensing data shown is generally centered on the metropolitan area of Houston, Texas.

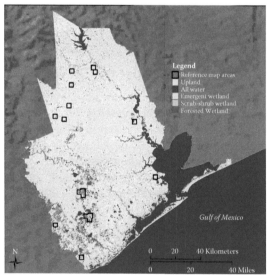

FIGURE 3.11
Eighteen stereopairs of aerial photographs (shown as individual or clustered sample areas) were used to check the accuracy of the satellite-based classification with the apparent presence of wetlands on the ground.

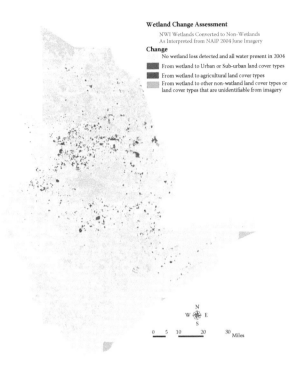

FIGURE 3.13

A wetland change analysis has been completed and is being used to improve the accuracy of the wetland map in the study area. The wetland change results describe the conversion of wetlands (per NWI maps) from the 1980s through summer 2004 (National Agriculture Imagery Program [NAIP] photography). An example of an area of wetland conversion that is currently "unidentifiable": a building that could be used for urban storage or nearby agriculture but could not be further clarified using the available remote sensing data or other geographic information.

FIGURE 4.2

Different surfaces reflect different amounts of light at various wavelengths so it is possible to identify general vegetational change from satellite measurements of reflected light as depicted here for the entirety of the U.S. and Canada Great Lakes Basin. (Courtesy of Burt Guindon, Canada Centre for Remote Sensing.)

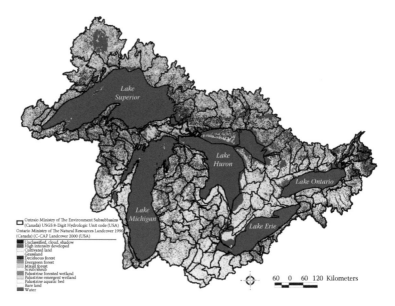

FIGURE 4.3
Land cover in the Great Lakes Basin as seen from space, using a detailed classification scheme that is a combination of the U.S. National Oceanographic and Atmospheric Administration's 2000s C-CAP land cover and Canada's Ontario Ministry of Natural Resource's 1990s land cover data sets.

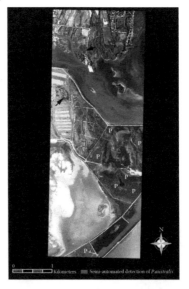

FIGURE 4.11
Results of a Spectral Angle Mapper (supervised) classification indicating likely areas of relatively homogeneous stands of *Phragmites australis* (solid blue), using PROBE-1 data and field-based ecological data. Black arrows show field-sampled patches of *Phragmites*. Areas of mapped *Phragmites* are overlaid on a natural-color image of Pointe Mouillee wetland complex (August 2001). The letter P indicates the general location of known areas of *Phragmites*, validated with aerial photographs, field notes, and 2002 accuracy assessment data.

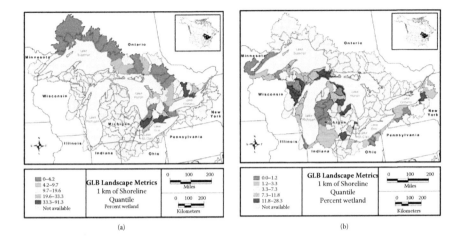

(a)

(b)

FIGURE 5.11

Extent of coastal wetlands is mapped here as percent of wetland within the 1 km coastal region among coastal watersheds in (a) Canada and (b) the United States. Percent coastal wetland is calculated by dividing the number of wetland land cover cells in the coastal region of each watershed (i.e., the reporting unit) by the total number of land cover cells in the reporting unit minus those cells classified as water. This measurement has potential for measuring and comparing wetland contribution among watersheds and may be used to indicate potential for wetland removal or reduction in the amount of pollutants entering the Great Lakes. The relative extent of coastal wetlands may also be developed into a quantitative indicator of habitat for a wide variety of plant and animal species.

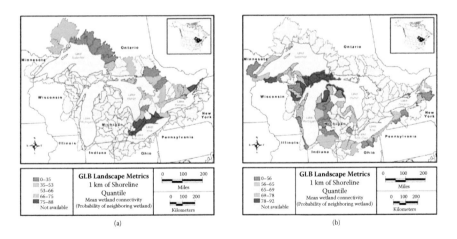

(a)

(b)

FIGURE 5.12

Mean wetland connectivity in a 1 km coastal region of the Great Lakes Basin (probability of neighboring wetland), which is the mean (for a reporting unit) probability of a wetland cell having a neighboring wetland cell, calculated using a moving 270 m^2 window (9 pixels × 9 pixels) across the GIS land cover data set. Because these analyses use two differing land cover data sets, results for (a) Canada and (b) the United States may not be directly comparable.

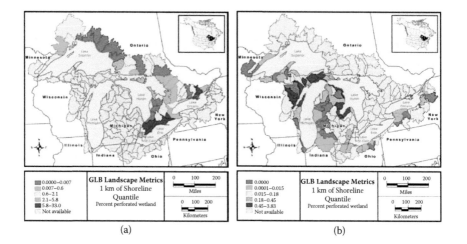

(a)   (b)

FIGURE 5.13
Percentage of perforated wetland, in a 1 km coastal region of the Great Lakes Basin, is calculated using a moving 270 m² window (9 pixels × 9 pixels) across the land cover, and generally indicates if center upland area(s) are present in a wetland. Because these analyses use two differing land cover data sets, results for (a) Canada and (b) the United States may not be directly comparable.

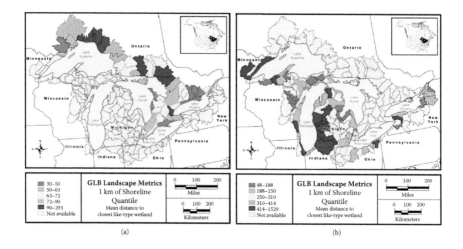

(a)   (b)

FIGURE 5.14
Mean distance to closest like-type wetland, in a 1 km coastal region of the Great Lakes Basin, is the mean minimum distance to closest wetland patch, for the 1 km shore area, within each hydrologic unit. Distances were measured from edge to edge and are reported in meters. This metric is useful in determining relative wetland habitat suitability at scales that are ecologically meaningful for specific plant and animal taxa. Because these analyses use two differing land cover data sets, results for (a) Canada and (b) the United States may not be directly comparable.

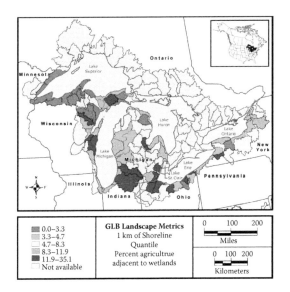

FIGURE 5.21
Percent agriculture adjacent to wetlands is mapped within a 1 km coastal region of the Great Lakes Basin (United States data only). The percentage of all agricultural land cover adjacent to wetlands is calculated by summing the total number of pasture and cropland land cover cells directly adjacent to wetland land cover cells in the reporting unit and dividing by wetland total area in the reporting unit.

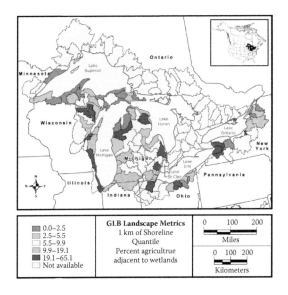

FIGURE 5.22
The percentage of urban land cover adjacent to wetlands is mapped within a 1 km coastal region of the Great Lakes Basin (United States data only). Percent urban is calculated by summing the total number of urban land cover cells directly adjacent to wetland land cover cells in the reporting unit and dividing by wetland total area in the reporting unit.

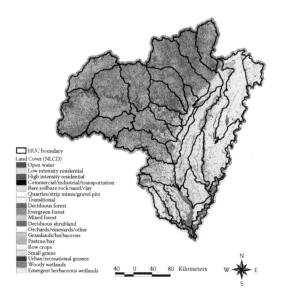

FIGURE 6.8

Land cover (National Land Cover Data [NLCD]) in the White River Watershed, showing subwatershed 8-digit HUC boundaries.

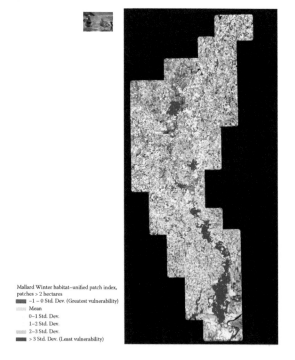

FIGURE 6.21 (a)

Lower White River Region (LWRR) (a) mallard duck, (b) black bear, and (c) wetland plant habitat vulnerability models (>2 ha) in terms of habitat patch size and shape.

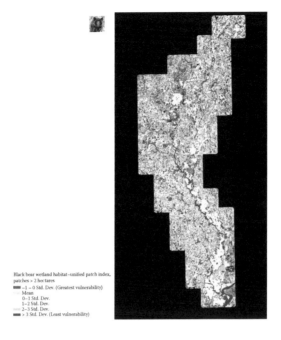

Black bear wetland habitat–unified patch index,
patches > 2 hectares
■ −1 – 0 Std. Dev. (Greatest vulnerability)
   Mean
   0–1 Std. Dev.
   1–2 Std. Dev.
   2–3 Std. Dev.
■ > 3 Std. Dev. (Least vulnerability)

FIGURE 6.21 (b)
Lower White River Region (LWRR) (a) mallard duck, (b) black bear, and (c) wetland plant habitat vulnerability models (>2 ha) in terms of habitat patch size and shape.

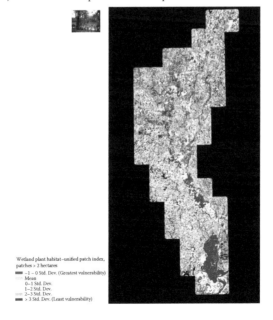

Wetland plant habitat–unified patch index,
patches > 2 hectares
■ −1 – 0 Std. Dev. (Greatest vulnerability)
   Mean
   0–1 Std. Dev.
   1–2 Std. Dev.
   2–3 Std. Dev.
■ > 3 Std. Dev. (Least vulnerability)

FIGURE 6.21 (c)
Lower White River Region (LWRR) (a) mallard duck, (b) black bear, and (c) wetland plant habitat vulnerability models (>2 ha) in terms of habitat patch size and shape.

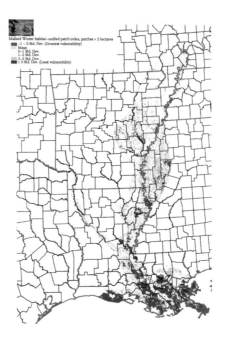

FIGURE 6.22 (a)

Mississippi Alluvial Valley Ecoregion (MAVE) (a) mallard duck, (b) black bear, and (c) wetland plant habitat vulnerability models (>2 ha) in terms of habitat patch size and shape.

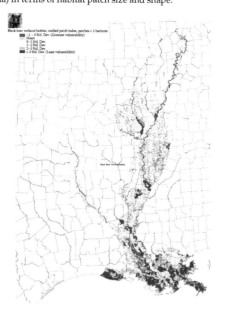

FIGURE 6.22 (b)

Mississippi Alluvial Valley Ecoregion (MAVE) (a) mallard duck, (b) black bear, and (c) wetland plant habitat vulnerability models (>2 ha) in terms of habitat patch size and shape.

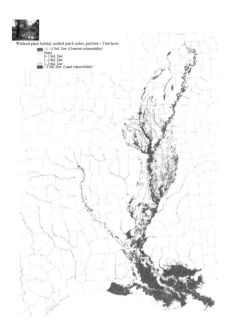

Wetland plant habitat, unified patch index, patches > 2 hectares
-1 – 0 Std. Dev. (Greatest vulnerability)
Mean
0–1 Std. Dev.
1–2 Std. Dev.
2–3 Std. Dev.
> 3 Std. Dev. (Least vulnerability)

FIGURE 6.22 (c)
Mississippi Alluvial Valley Ecoregion (MAVE) (a) mallard duck, (b) black bear, and (c) wetland plant habitat vulnerability models (>2 ha) in terms of habitat patch size and shape.

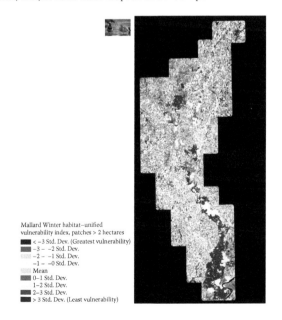

Mallard Winter habitat–unified
vulnerability index, patches > 2 hectares
< –3 Std. Dev. (Greatest vulnerability)
–3 – –2 Std. Dev.
–2 – –1 Std. Dev.
–1 – –0 Std. Dev.
Mean
0–1 Std. Dev.
1–2 Std. Dev.
2–3 Std. Dev.
> 3 Std. Dev. (Least vulnerability)

FIGURE 6.23 (a)
Lower White River Region (LWRR) (a) mallard duck, (b) black bear, and (c) wetland plant habitat vulnerability models (>2 ha) in terms of human-induced disturbance factors.

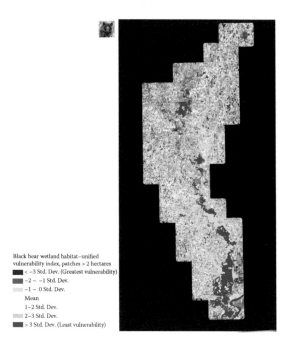

Black bear wetland habitat–unified
vulnerability index, patches > 2 hectares
█ < −3 Std. Dev. (Greatest vulnerability)
█ −2 − −1 Std. Dev.
░ −1 − 0 Std. Dev.
  Mean
  1–2 Std. Dev.
░ 2–3 Std. Dev.
█ > 3 Std. Dev. (Least vulnerability)

FIGURE 6.23 (b)

Lower White River Region (LWRR) (a) mallard duck, (b) black bear, and (c) wetland plant habitat vulnerability models (>2 ha) in terms of human-induced disturbance factors.

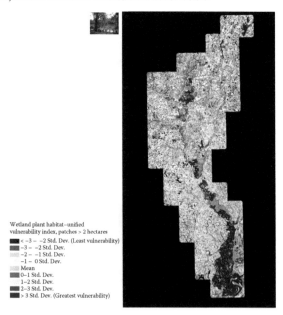

Wetland plant habitat–unified
vulnerability index, patches > 2 hectares
█ < −3 − −2 Std. Dev. (Least vulnerability)
█ −3 − −2 Std. Dev.
░ −2 − −1 Std. Dev.
  −1 − 0 Std. Dev.
░ Mean
█ 0–1 Std. Dev.
  1–2 Std. Dev.
█ 2–3 Std. Dev.
█ > 3 Std. Dev. (Greatest vulnerability)

FIGURE 6.23 (c)

Lower White River Region (LWRR) (a) mallard duck, (b) black bear, and (c) wetland plant habitat vulnerability models (>2 ha) in terms of human-induced disturbance factors.

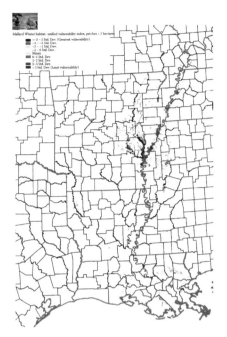

FIGURE 6.24 (a)
Mississippi Alluvial Valley Ecoregion (MAVE) (a) mallard duck, (b) black bear, and (c) wetland plant habitat vulnerability models (>2 ha) in terms of human-induced disturbance factors.

FIGURE 6.24 (b)
Mississippi Alluvial Valley Ecoregion (MAVE) (a) mallard duck, (b) black bear, and (c) wetland plant habitat vulnerability models (>2 ha) in terms of human-induced disturbance factors.

FIGURE 6.24 (c)
Mississippi Alluvial Valley Ecoregion (MAVE) (a) mallard duck, (b) black bear, and (c) wetland plant habitat vulnerability models (>2 ha) in terms of human-induced disturbance factors.

FIGURE 6.26 (a)
(a) Single near-infrared band image of the Lower White River Region with selected towns indicated with blue arrows. Wildlife refuge zones are indicated by white polygons. (b) Same image displayed in a three-band false-color infrared composite image. (c) Same image displayed in a three-band false-color "enhanced vegetation" infrared composite image.

FIGURE 6.26 (b)

(a) Single near-infrared band image of the Lower White River Region with selected towns indicated with blue arrows. Wildlife refuge zones are indicated by white polygons. (b) Same image displayed in a three-band false-color infrared composite image. (c) Same image displayed in a three-band false-color "enhanced vegetation" infrared composite image.

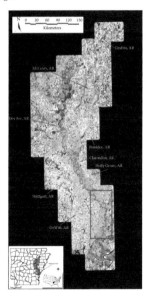

FIGURE 6.26 (c)

(a) Single near-infrared band image of the Lower White River Region with selected towns indicated with blue arrows. Wildlife refuge zones are indicated by white polygons. (b) Same image displayed in a three-band false-color infrared composite image. (c) Same image displayed in a three-band false-color "enhanced vegetation" infrared composite image.

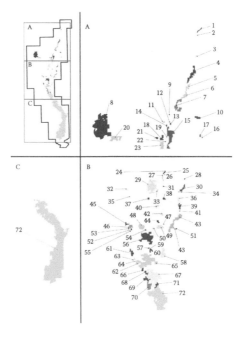

FIGURE 6.27
Seventy-two federal refuge zones in the Lower White River Region were compared using mallard duck winter habitat vulnerability models.

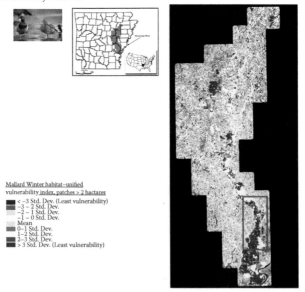

Mallard Winter habitat–unified
vulnerability index, patches > 2 hactares
■ < −3 Std. Dev. (Least vulnerability)
■ −3 – 2 Std. Dev.
  −2 – 1 Std. Dev.
  −1 – 0 Std. Dev.
  Mean
■ 0–1 Std. Dev.
■ 1–2 Std. Dev.
■ 2–3 Std. Dev.
■ > 3 Std. Dev. (Least vulnerability)

FIGURE 6.28
The unified vulnerability index was applied to a hypothetical landscape where change in the riparian wetland hydrology has occurred. Red rectangle indicates the portion of the South Unit in the White River National Wildlife Refuge where landscape change is hypothesized.

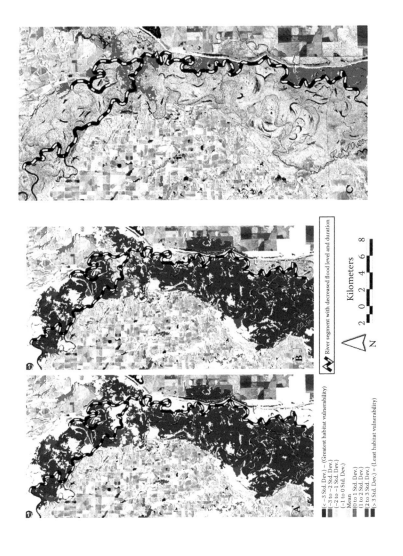

FIGURE 6.29

(a) Mallard duck winter habitat vulnerability under current hydrologic conditions in the South Unit of the White River National Wildlife Refuge; (b) loss of mallard duck winter habitat under hypothetical decrease in flood stage and duration on the White River; (c) enlarged view of predicted net loss of mallard duck winter habitat under hypothetical decrease in flood stage and duration on the White River.

6

Determining Ecological Functions of Wetlands with Landscape Characterization

Gathering Information

There are a variety of information needs in landscape characterization and wetland ecosystem management. This is due in part to the variety of issues to be addressed. The issues include the traditional examples and related questions that develop from analysis of the problems.

The traditional issues and questions come from the operational information needs of conservation, environmental groups, and government. These specific data needs may include characteristics such as wetland feature type and size, wetland exposure and risk, conditions of a site and facilities or resources, management activities, and other information.

The domain of these issues includes new requirements beyond the traditional and the questions they engender. New issues and questions most often stem from regulatory requirements. The evolution of the requirements comes from a variety of federal, state, and local agencies. These new types of requirements spawn different questions and also the need for data. New information needs include data on human factors, data for identification and management of wetland and related ecosystems, cultural features, databases to support management operations and maintenance, and data to support public query and response systems.

Management Considerations and Maximizing Tools

Many questions related to assessing landscape condition and measuring wetland functions may involve encroachment in and around wetland properties. There are a number of imaging and GIS methods to identify the location and the probability that an activity or exposure has occurred. These are in addition to traditional field measures.

FIGURE 6.1
From the ground view, the same tonal differences of soil and vegetation materials can be seen. Note that soils can range from dark tones with organic matter and moisture to lighter tones. Soils have a medium and steady reflectance in the visible and near-infrared resulting in generally light tones but here are quenched by organic matter or moisture. Conversely, in this winter season scene, plant residue is without chlorophyll and is uniformly light toned. Pictured here is a farm field and wetland grasslike plant residue in South Dakota.

Typically, an activity can be identified by the change in gray tone or color on individual images. A comparison with additional images (such as ground photos) allows further assessment of conditions, and identifies whether there has been activity or a disruption (Figure 6.1).

Much of the remote capability for identification employs the interpretation of image products over time. The use of multiple images can facilitate a thorough general evaluation of conditions from one timeframe to another.

Once a disturbance or exposure has been identified it can be checked in the field. This approach allows for multiple benefits including efficient use of personnel by guiding them to a probable activity, improving safety as well as providing detailed field-based information. The approach also represents a good application of capabilities, technologies supplying location and identification data, and field personnel checking for activities or exposure.

As described in several instances throughout the book, remote sensor data and GIS technologies also assist in the prioritization of field crew work. Information can aid in the dispatch of field crews and in the facilitation of the crew's evaluations of a site. The imaging remote sensor data and GIS supply locations for field checks, and they supply data in the form of map sheets, images of features, data on wetland characteristics, and other appropriate data if available.

Methods such as detection of change in wetlands and other land cover areas hold promise. They can be particularly cost effective as compared to the relative costs of traditional methods of field work, and procurement and use of aerial photography (Falker and Morgan 2001). Benefits include the use of satellite and aerial remote sensor data for detection of change, production and use of orthophoto-like maps, data inputs to GIS systems, monitoring of resource integrity, evaluation of areas following floods or earthquakes, identification and storage of data concerning cultural and natural resource features, and the production of data sheets from GIS for planning and management.

A great deal of research has gone into the development of change detection procedures (Lunetta et al. 1993; Lunetta and Elvidge 1998; Lyon et al. 1998). To optimize the process, applicable procedures utilize digital imaging data from two or more dates. The data may come from the same aerial or satellite sensor, or may be from different sources. The data sets are evaluated for change over time. Initially, one data set can be preprocessed to the characteristics of the second data set. The two data sets are also placed in a common mapping framework or mapping projection. Completion of these steps facilitates the comparisons of images.

Planning Considerations

The benefits to wetlands of a remote measurement and GIS systems are multifold. Of particular value are remote imagery data and spatial data systems in the planning phase of a project (Figure 6.2). For planning, the data are useful in the optimization of avoidance of wetlands, the optimization of management activities, relative economics, and other important factors.

The use of remote sensor data and spatial information systems like GIS are valuable in reporting information. The planning phase of a project or ongoing management will often require information for permitting related to the National Environmental Policy Act (NEPA), and general reporting to government entities, the public, and management personnel. Remote sensor and spatial (GIS) data have definitively proven to be invaluable for stakeholders and for planning purposes.

A recent concern by the public, industry, and governmental agencies is the extent and variety of wetlands on the continental and global scale as these elements relate to greenhouse gasses and associated global climate considerations. To address wetland resources at either the local or global scale requires identification, inventory, and monitoring to supply information for informed decisions.

Over the past ten years, improvements in computing, imaging, and spatial analysis capabilities have been the primary methods for meeting the needs of agencies and companies for operations and management, or planning information (Figure 6.3). These advances in wetland landscape characterization

FIGURE 6.2
An aerial view of a riparian and wetland area with little water in the channel and growing season plant material that is dark toned. The bare soils revealed by earthmoving, which is generally indicated by the lighter and somewhat irregular tones, as opposed to dark-toned plant material in the fields and in the channel. Also note the machinery near the soil excavation surfaces and the clearing of the channel way.

FIGURE 6.3
Pictured are tidal mudflats that have wetland characteristics. Note the new lands look of incised drainage ways and dark-toned soils. Often, small clues about site characteristics can indicate human activities, faintly indicated in this ground-based image by the distant power transmission lines (upper right).

also allow governmental conservation and environmental groups and others to share data.

Features of Interest

Much of the analysis of remote sensor data revolves around the presence or absence of features (Figure 6.4), such as flooded areas, buildings, or vegetation. The features of interest in wetlands include topography, soils, and hydrology. Additional features of interest are those that indicate favorable conditions for preservation of wetlands or the construction of mitigation wetlands, supplying valuable data for maintenance and monitoring of wetlands and related ecosystems.

Adjacent locations of features are important landscape characteristics to help those who manage wetlands. Basic data requirements usually include information about the surrounding land uses or land cover types. The use of remote sensor digital data and GIS technologies can be used along with information about the associated wetlands themselves. Appropriate data includes location of the public and their activities, location of built or

FIGURE 6.4
A close-up, low-altitude image of a drainage way that has been worked. Note the light tone of the road materials hugging the north shore, the pattern of the fields where relatively dark-toned plant materials create a pattern of crop structure against the lighter-toned soils. Here the trees can be seen by the dark-toned plant material and shadow as well as the lighter-toned trunks that appear.

constructed features, and the location of features with respect to wetlands and facilities.

Wetland Risks and Impacts

There are a number of activities or events that can cause risk to wetlands. Many of these activities or events can be detected using a GIS and can be identified from indicators or surrogate variables, (e.g., the size or shape of habitat patches, or land cover type) and often lead to early identification and remedial actions.

The monitoring of wetland risks and impacts can be accomplished through the use of aerial photographs (Figure 6.5) and satellite remote sensor measurements. The identification of an activity may involve changes in land cover, and associated plant or animal habitat conditions, or water resources conditions.

In certain areas, the conditions of wetlands are much different from other areas and they present problems unique to the location or application (e.g., habitat risk or water quality risk). Often these conditional differences are the result of a combination of geological or hydrological characteristics. Additionally, due to regulatory considerations, the human activities or

FIGURE 6.5
In the semiarid west one can see the use of strip cropping to hold over biannual soil moisture for successful alternate year cropping. The light-tone strips are soils and the dark toned are crops. Juxtaposed are the untamed lower lands and drainage ways that show plant materials in irregular dark tones and very dark vegetated areas that show some indications of being wetlands, typical of the western United States.

features in or adjacent to properties may need to be identified and monitored. A major current interest is the identification and location of human activities that constitute environmental risk near or in wetland ecosystems.

Case Study: Habitat Risk Assessments in the Lower Mississippi River Basin Using Metrics and Indicators

This case study makes use of ecological indicators, landscape metrics, and landscape indicators. The comprehensive case study provides a definitive illustration of the many factors involved in using metrics and indicators. For the purposes of this case study an *ecological indicator* is a sample measurement of an ecological resource (Bromberg 1990; Hunsaker and Carpenter 1990), typically from field sampling. When measured at a relatively broad *landscape scale* (Forman 1995), *landscape metrics* (e.g., percent forest cover) that are characteristic of the environment, as measured by a sufficient sample size of ecological indicators, can provide quantitative information about ecological resources at broad scales and are referred to as *landscape indicators*.

Landscape metrics in this case study for habitat analyses (Table 6.1) and available for water quality analyses (see Lopez et al., 2008 for water quality analyses; Table 6.2) are similar to those typically used to correlate with ecological indicators at several scales. Utilization of the correlation technique is fairly commonplace at relatively broad scales (e.g., Riitters et al. 1995; Jones et al. 2000; Jones et al. 2001), such as the entire Mississippi River Valley; at moderate scales (e.g., van der Valk and Davis 1980; Nagasaka and Nakamura 1999; Roth et al. 1996; Fauth et al. 2000; Lopez et al. 2002; Lopez and Fennessy 2002), such as a wildlife refuge; and at single-site or mesocosm scales (e.g., Peterjohn and Correll 1984; Murkin and Kadlec 1986; Ehrenfeld and Schneider 1991; Willis and Mitsch 1995; McIntyre and Wiens 1999a, 1999b; Luoto 2000), such as individual units of a wildlife refuge. The combined use of observed correlations, GIS mapping techniques, and statistical analysis techniques for habitat assessments (the focus of this case study) across the Lower White River Region (LWRR), Mississippi Alluvial Valley Ecoregion (MAVE), and a portion of the White River case study areas facilitated the determination of land cover gradients and ecological vulnerability at each scale, aided by geospatial analysis techniques (after Scott et al. 1993; Jones et al. 1997).

This case study uses different landscape gradients. Each gradient is a range of a condition, which is observed across a selected landscape unit (e.g., across the LWRR) or among reporting units (e.g., among wildlife refuges). Thus the two important selection criteria for study areas in the landscape gradient assessment were (1) a sufficient range of conditions along each landscape

TABLE 6.1

Habitat Parameters Used to Produce Models in Study Areas

Habitat Vulnerability Parameter	Interpretation and Calculation
Wetland habitat patch total area for patches greater than 2 ha	Smaller patches are relatively less likely to rebound from disturbances (i.e., are more likely to be fragmented or destroyed after changes in hydrology, destruction of vegetation, or the establishment of opportunistic flora and fauna) than larger patches.
Wetland habitat patch total perimeter length for patches greater than 2 ha	Patches with shorter perimeters are relatively less likely to rebound from disturbances than patches with longer perimeters. That is, shorter perimeter length values indicate that a patch has a greater likelihood of fragmentation or loss as a result of environmental change, such as changes in hydrology, destruction of vegetation, or the establishment of opportunistic flora and fauna.
Wetland habitat patch interior to edge ratio for patches greater than 2 ha	Patches with a smaller interior-to-edge ratio are relatively less likely to rebound from disturbances than patches with a larger interior-to-edge ratio. That is, smaller ratio values indicate that a patch has a greater likelihood of fragmentation or loss as a result of environmental change, such as changes in hydrology, destruction of vegetation, or the establishment of opportunistic flora and fauna. Calculation: Area/Perimeter, where Perimeter is the patch perimeter and Area is the patch area.
Wetland habitat patch sinuosity index for patches greater than 2 ha	Patches with a smaller sinuosity index are less winding or convoluted in shape, thus are relatively less likely to rebound from disturbances than patches with a larger sinuosity index. That is, smaller index values indicate that a patch has a greater likelihood of fragmentation or loss as a result of environmental change, such as changes in hydrology, destruction of vegetation, or the establishment of opportunistic flora and fauna. Calculation: $Perimeter/\{2 \times \pi \times [\sqrt{(Area/p)}]\}$, where Perimeter is the patch perimeter and Area is the patch area (after Bosch 1978 and Davis 1986).
Wetland habitat patch circularity index for patches greater than 2 ha	Patches with a smaller circularity index are more circlelike in shape, thus are relatively less likely to rebound from disturbances than patches with a larger circularity index. That is, smaller index values indicate that a patch has a greater likelihood of fragmentation or loss as a result of environmental change, such as changes in hydrology, destruction of vegetation, or the establishment of opportunistic flora and fauna. Calculation: $\{p \times [Perimeter/(2 \times p)]2\}/Area$, where Perimeter is the patch perimeter and Area is the patch area (after Stoddart 1965 and Unwin 1981).

(continued)

TABLE 6.1 (CONTINUED)

Habitat Parameters Used to Produce Models in Study Areas

Habitat Vulnerability Parameter	Interpretation and Calculation
Wetland habitat unified patch index for patches greater than 2 ha	Patches with a smaller unified patch index are relatively less likely to rebound from disturbances than patches with a larger unified patch index. That is, smaller index values indicate that a patch has a greater likelihood of fragmentation or loss as a result of environmental change, such as changes in hydrology, destruction of vegetation, or the establishment of opportunistic flora and fauna. Calculation: Pier × CIRC_ind × SIN_ind, where Pier is the patch interior to edge ratio, CIRC_ind is the patch circularity index, and SIN_ind is the patch sinuosity index.
Wetland habitat patch total road length for patches greater than 2 ha	Patches with greater total road length are relatively less likely to rebound from disturbances than patches with a lesser total road length. That is, greater length values indicate that a patch has a greater likelihood of fragmentation or loss as a result of environmental change, such as changes in hydrology, destruction of vegetation, or the establishment of opportunistic flora and fauna. The increased presence of roads may also bring about the aforementioned disturbances.
Wetland habitat patch total road density for patches greater than 2 ha	Patches with greater total road density are relatively less likely to rebound from disturbances than patches with a lesser total road density. That is, greater density values indicate that a patch has a greater likelihood of fragmentation or loss as a result of environmental change, such as changes in hydrology, destruction of vegetation, or the establishment of opportunistic flora and fauna. The increased presence of roads may also bring about the aforementioned disturbances.
Wetland habitat patch road index for patches greater than 2 ha	Patches with a greater road index are relatively less likely to rebound from disturbances than patches with a lesser road index. That is, greater index values indicate that a patch has a greater likelihood of fragmentation or loss as a result of environmental change, such as changes in hydrology, destruction of vegetation, or the establishment of opportunistic flora and fauna. The increased presence of roads may also bring about the aforementioned disturbances. Calculation: Rddens + Rdlen, where Rddens is the number of roads per patch area and Rdlen is the total length of road per patch area.

(continued)

TABLE 6.1 (CONTINUED)

Habitat Parameters Used to Produce Models in Study Areas

Habitat Vulnerability Parameter	Interpretation and Calculation
Wetland habitat patch human population density in 1990 for patches greater than 2 ha	Patches that reside in census block groups with a greater population density are relatively less likely to rebound from disturbances than patches that reside in areas of lesser population density. That is, greater density values indicate that a patch has a greater likelihood of fragmentation or loss as a result of environmental change, such as changes in hydrology, destruction of vegetation, or the establishment of opportunistic flora and fauna. The increased presence of residents near patches may also bring about the aforementioned disturbances.
Wetland habitat patch human population density in 2011 for patches greater than 2 ha	Patches that reside in census block groups with a greater population density in the future are relatively less likely to rebound from disturbances than patches that reside in areas of lesser population density in the future. That is, greater density values indicate that a patch has a greater likelihood of fragmentation or loss as a result of environmental change, such as changes in hydrology, destruction of vegetation, or the establishment of opportunistic flora and fauna. The increased presence of residents near patches may also bring about the aforementioned disturbances.
Wetland habitat patch human population density change from 1990 to 2011 for patches greater than 2 ha	Patches that reside in census tracts with a greater increase in population density are relatively less likely to rebound from disturbances than patches that reside in areas of lesser population density change. That is, greater (predicted) increases in human population density indicate that a patch has a greater likelihood of fragmentation or loss as a result of environmental change, such as changes in hydrology, destruction of vegetation, or the establishment of opportunistic flora and fauna. The increased density of human population near patches may also bring about the aforementioned disturbances. Calculation: popden2 – popden1, where popden2 is the 2011 population density and popden1 is the population density in 1990.

(continued)

TABLE 6.1 (CONTINUED)

Habitat Parameters Used to Produce Models in Study Areas

Habitat Vulnerability Parameter	Interpretation and Calculation
Wetland habitat patch unified human index for patches greater than 2 ha	Patches with a greater unified human index are relatively less likely to rebound from disturbances than patches with a smaller unified human index. That is, greater index values indicate that a patch has a greater likelihood of fragmentation or loss as a result of environmental change, such as changes in hydrology, destruction of vegetation, or the establishment of opportunistic flora and fauna. The increased presence of roads and growth of residential areas near patches may also bring about the aforementioned disturbances. Calculation: $[(popdenchg90\_11) + C] + [\sqrt{(rdden + len)}]$, where popdenchg90_11 is the population density change between 1990 and 2011, C is a normalization quantity to ensure that the net change value is positive, and rdden + len is the road index.
Wetland habitat patch unified vulnerability index for patches greater than 2 ha	Patches with a smaller unified vulnerability index are less likely to rebound from disturbances than patches with a larger unified vulnerability index. That is, smaller index values indicate that a patch has a greater likelihood of fragmentation or loss as a result of environmental change, such as changes in hydrology, destruction of vegetation, or the establishment of opportunistic flora and fauna. Calculation: $[1/(\sqrt{UPI})] + (\sqrt{UHI})$, where UPI is the patch unified patch index and UHI is the patch unified human index.

gradient of interest within a study area and (2) a sufficient number of sites to compare among reporting units (Green 1979; Karr and Chu 1997).

The case study began with initial visual analysis of available GIS and remote sensing data (Table 6.3, Table 6.4, and Table 6.5 [for reference only]), and meetings with local experts about site-based information and data availability from multiple partners at the Arkansas Game and Fish Commission, the Arkansas Natural Heritage Commission, the Arkansas Soil and Water Conservation Commission, The Nature Conservancy, U.S. Army Corps of Engineers (USACE), U.S. Environmental Protection Agency (USEPA), and U.S. Fish and Wildlife Service (USFWS).

There is an inherent trade-off between conducting site-based studies, which are limited by a lack of contextual information about the surrounding landscape, and landscape-scale studies, which are limited by a lack of detailed information about small areas. Therefore, although landscape-scale assessments are founded on the ecological principles of site-specific studies,

TABLE 6.2

Water Quality Metrics Available for Use in Study Areas (after Lopez
et al. 2008)

Water Quality Vulnerability Metric
Largest forest patch proportion of HUC
Mean forest patch area
Largest forest patch area
Forest patch number
Forest patch density
Percent streams within 30 m of roads
Percent total agriculture on slopes greater than 3%
Percent total agriculture within entire HUC
Percent crop agriculture within entire HUC
Percent pasture within entire HUC
Percent forest within entire HUC
Percent wetland within entire HUC
Percent natural land cover within entire HUC
Percent total agriculture within a 300 m riparian zone at 30 m increments
Percent crop agriculture within a 300 m riparian zone at 30 m increments
Percent pasture within a 300 m riparian zone at 30 m increments
Percent forest within a 300 m riparian zone at 30 m increments
Percent wetland within a 300 m riparian zone at 30 m increments
Percent natural land cover within a 300 m riparian zone at 30 m increments

additional information may improve outputs and should be considered as
part of a complete process. Case study models may also be limited by the
general lack of detailed information about small habitat areas. Thus the
models are intended as a tool for large areas, which would otherwise be
impractical to assess in the field. Ideally, these GIS-based models would be
used in combination with, and guide or supplement, detailed field investiga-
tions (see Chapter 4).

Study Area Background

The LWRR and the MAVE study areas demonstrate the assessment of landscape conditions and measurement of wetland conditions (Figure 6.6). The
White River begins in the mountainous northwestern region of Arkansas,
flows through southwestern Missouri, reenters north central Arkansas and
flows down from the Ozark Mountains into Arkansas's agricultural plain,
where it meanders to its confluence with the Mississippi River (Figure 6.7). The
catchment area of the White River Watershed (WRW) (Figure 6.6) extends from
the Fayetteville, Arkansas, in the western Ozark Mountains to the Mississippi
River, and drains from a wide range of landscapes containing farmland,
upland forests, wetlands, lakes, streams, and urban areas (Figure 6.8).

TABLE 6.3

GIS Data Sets Used to Derive Mallard Duck, Black Bear, Least Tern, and Wetland Plant Habitat Suitability and Vulnerability in Lower White River Region

Data Used	Derived Land Cover or Land Use	Reference or Source	Relevant Internet URL
Mississippi Alluvial Valley of Arkansas Landuse/Landcover (MAVA-LULC)	Agriculture crop cover	Gorham 1999	www.cast.uark.edu/ local/lulc/
National Hydrography Dataset (NHD) v. 1	Surface water location	USGS and USEPA, 1999	http://nhd.usgs. gov/
National Wetland Inventory (NWI)	Wetland cover and hydrology type	USFWS, various	http://www.fws. gov/wetlands/ index.html
Arkansas GAP Program (AR-GAP)	Vegetation community cover, vegetation taxa cover, surface water location, agriculture cover, urban cover	Smith et al. 1998	http://web.cast.uark. edu/gap/
Historical Forest Cover	1950s forest cover	Llewellyn et al. 1996	
U.S. Census Bureau Statistics	Current and future estimates of human population	Applied Geographic Solutions, 2001	http://www. appliedgeographic. com/ags_data_ software.html
Wessex Streets v. 7.0	Road length and density	Geographic Data Technology, 1999	

There are seven major dams that maintain large reservoirs for flood prevention and recreation along the White River and a national scenic river. Along the banks of the White River and its tributaries there are two national wildlife refuges, two national forests, and one national scenic river. The Cache River (a tributary to the White River) and its wetlands have been designated as Ramsar Wetlands of International Importance (Ramsar 2002), along with 1,235 other wetlands around the world. The LWRR contains most of the White River channel that flows through Arkansas' agricultural plain, which is a region currently dominated by row-crop agriculture and rice fields (Figure 6.8) but also contains a large proportion of the Mississippi River Valley's last remaining bottomland hardwood swamps (Dahl 1990).

The LWRR provides suitable habitat for the largest winter concentration of mallard ducks (*Anas platyrhynchos*) and other ducks in North America, provides necessary habitat for recovering populations of black bears (*Ursus americanus*), and provides some of the last remaining habitat for many (some unique) wetland plants in the region (Carreker 1985; Allen 1987; Rogers and

TABLE 6.4

GIS Data Sets Used to Derive Mallard Duck, Black Bear, Least Tern, and Wetland Plant Habitat Suitability and Vulnerability in Mississippi Alluvial Valley Ecoregion

Data Used	Derived Land Cover or Land Use	Reference or Source	Relevant Internet URL
National Land Cover Data (NLCD)	Open water, residential/ commercial/industrial, barren, agriculture, forest, shrubland, grassland/herbaceous, woody wetland, and emergent herbaceous wetland cover	Vogelmann et al. 2001	http://landcover.usgs.gov/natllandcover.html
National Hydrography Dataset (NHD) v. 1	Surface water location	USGS and USEPA, 1999	http://nhd.usgs.gov/
U.S. Census Bureau Statistics	Current and Future Estimates of Human Population	Applied Geographic Solutions, 2001	www.appliedgeographic.com/datadescripts. htm#censussf1
Wessex Streets v. 7.0	Road length and density	Geographic Data Technology, 1999	www.geographic.com/home/index.cfm

TABLE 6.5

GIS Data Sets Available to Derive Water Quality Vulnerability Models in the Study Area (after Lopez et al. 2008)

Data Used	Derived Land Cover or Land Use	Reference or Source	Relevant Internet URL
National Land Cover Data (NLCD)	Open water, residential/ commercial/industrial, barren, agriculture, forest, shrubland, grassland/ herbaceous, woody wetland, and emergent herbaceous wetland cover	Vogelmann et al. 2001	http://landcover. usgs.gov/ natllandcover.html
National Hydrography Dataset (NHD) v. 1	Surface water location	USGS and USEPA, 1999	http://nhd.usgs. gov/
Wessex Streets v. 7.0	Road length and density	Geographic Data Technology, 1999	http://www. appliedgeographic. com/
National Elevation Dataset (NED)	Topographic slope	Gesch et al. 2002	http://gisdata.usgs. net/NED/default. asp

Allen 1987). The White River aquatic ecosystems also support an important riverine fishery, including sturgeon (*Scaphirhynchus albus* and *S. platorynchus*) and paddlefish (*Polyodon spathula*), and aquatic plant communities within the bottomland hardwood swamps, which represent some of the most biologically diverse and productive ecosystems of the world (Mitsch and Gosselink 1993). The White River and the surrounding landscape also contain valuable resources for the people living and working in the region because they provide the ecosystem services of irrigation water, drinking water, flood control, transportation for agricultural commerce, recreation, commercial shelling and fishing products, tourism, and attractions related to the famed (and once thought to be extinct) ivory-billed woodpecker, rediscovered in 2004 within the study site (http://www.fus.gov/ivorybill/).

Although the predominant land cover in the LWRR and MAVE is currently agriculture, forest land cover predominates in the Ozark Mountain region of the WRW. Closer inspection of the land cover throughout the WRW reveals that the landscape is a mosaic of many different discernible land cover types, such as dry upland forests, wetlands, human-built and populated areas, pastures, and row-cropped land. A mixture of land cover occurs, for example, as a gradient (i.e., a gradual change in the relative proportions) of land cover types from the northwestern corner of the WRW to the southeastern corner of the WRW. Similar landscape gradients to this are used in here to predict habitat suitability (i.e., the applicability of land cover to organismal

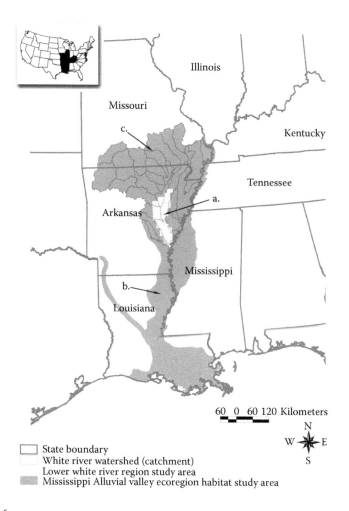

FIGURE 6.6
Orientation map of Lower White River Region (LWRR), Mississippi Alluvial Valley Ecoregion (MAVE), and White River Watershed (WRW). Inset shows the states that intersect the MAVE study area.

requirements) and vulnerability (i.e., the risk of loss or damage of suitable habitat) for mallard ducks, black bears, and wetland plants. Vulnerability predictions were performed with the use of currently available land cover data for the LWRR and the MAVE, using habitat boundaries that are associated with ecoregions (Omernik 1987).

Historically, the LWRR, the MAVE, and the WRW have all undergone substantial alterations in land cover (Dahl 1990), particularly the conversion of wetlands (Figure 6.9) to agricultural land (Figure 6.10). Consequently, there has been tremendous biological and hydrologic change throughout the landscape of the region (Dahl 1990; Mitsch and Gosselink 1993), particularly in

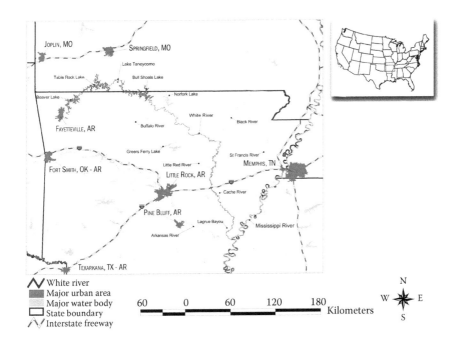

FIGURE 6.7
Geographic overview of the Lower White River Region (LWRR) and White River Watershed (WRW).

riparian wetlands (Figure 6.11 and Figure 6.12). Although the majority of wetland losses in the region occurred prior to the 1970s, the trend has continued in Arkansas, Mississippi, and Louisiana as a result of wetland conversion (Johnston 1989; Dahl and Johnson 1991; The Nature Conservancy 1992; Kress et al. 1996; Heggem et al. 2000; National Resources Conservation Service 2000). In particular, an estimated 70% of Arkansas' wetlands have been converted to other land cover types since the late nineteenth century (Dahl 1990), a change of approximately 28,000 km^2 of wetland, with greater than 4,000 km^2 of wetland loss occurring in Arkansas during the first half of the twentieth century (Shaw and Fredine 1956). The ongoing losses of wetlands in the region have been positively correlated with regional and local losses of biological diversity (Gosselink and Turner 1978; Ewel 1990; Kilgor and Baker 1996; Smith 1996; Wakeley and Roberts 1996); an increase in frequency, severity, and duration of flood events (Hopkinson and Day 1980a, 1980b; Brown 1984); and the degradation of downstream water quality (Kitchens et al. 1975; Day et al. 1977; Hupp and Morris 1990; Hupp and Bazemore 1993; DeLaune et al. 1996; Dortch 1996; Kleiss 1996; Long and Nestler 1996; Walton et al. 1996a, 1996b; Wilber et al. 1996). Therefore the information provides an excellent and deep understanding of the landscape conditions, which adds the necessary context for analysis.

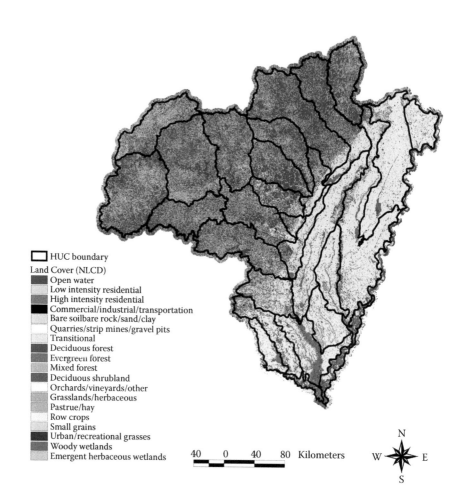

HUC boundary
Land Cover (NLCD)
Open water
Low intensity residential
High intensity residential
Commercial/industrial/transportation
Bare soilbare rock/sand/clay
Quarries/strip mines/gravel pits
Transitional
Deciduous forest
Evergreen forest
Mixed forest
Deciduous shrubland
Orchards/vineyards/other
Grasslands/herbaceous
Pastrue/hay
Row crops
Small grains
Urban/recreational grasses
Woody wetlands
Emergent herbaceous wetlands

40   0   40   80 Kilometers

N
W   E
S

FIGURE 6.8
(See color insert.) Land cover (National Land Cover Data [NLCD]) in the White River Watershed, showing subwatershed 8-digit HUC boundaries.

The influence of landscape condition on vegetation is intimately linked with those organisms that rely on the vegetation for forage, shelter, and areas to breed. These conditions are closely related to the success of those organisms that may be the focus of decision making at the local, regional, or national scales of analysis.

This case study demonstrates how land cover configuration (e.g., size and shape of habitat *patches*) may influence habitat condition, which can be used for land use planning. Future development in the vicinity of the White River and within the MAVE has great potential for altering wetland and aquatic ecosystem habitat conditions. This may include oil and gas production, or other activities driven by societal needs. There are several land development projects and human activities in the vicinity of the White River that are

FIGURE 6.9
Wetland disturbances are the result of a number of different human activities.

planned or ongoing, each of which has the potential to damage or destroy a substantial portion of the remaining wetlands in the area. In general, planned or ongoing development activities in the area include construction of dikes or channel modifications to increase river flow rates; agricultural irrigation projects that involve increasing the removal of surface water from the White River to supplement agriculture irrigation shortages in dry years (Jehl 2002); and modification of reservoir release schedules.

FIGURE 6.10
Diesel generator pumping groundwater for the irrigation of row crops in Arkansas County, Arkansas.

FIGURE 6.11
An example of riparian vegetation destruction and wetland hydrologic alteration.

FIGURE 6.12
Alterations in river and upland hydrologic patterns and associated structures can impact riparian wetlands.

Additional contextual data about watershed conditions were accessed from the USEPA at http://cfpub.epa.gov/surf/locate/index.cfm (accessed July 21, 2012) and used to bolster the landscape condition information determined from landscape analyses. There are increased emphases highlighted currently that focus on both the biotic and abiotic conditions of the landscape, and the matrix of these conditions that exist in wetlands should be accounted for, ideally, to convey an integrated view of landscape conditions and functions, linking wetlands with open water and upland ecosystems as well as societal elements.

Identification of Habitat for Conservation or Mitigation

Identification of habitat for species has been a consideration in siting constructed wetlands or selecting property for acquisition for mitigation of wetlands. Much like any project, a checklist of items is necessary for adequate analyses and description of wetland conditions. New items for the checklist may include presence or absence of plant and animal species of interest; presence or absence of general or jurisdictional wetlands; and presence or absence of cultural resources.

One GIS method to address plant and animal species is to take advantage of governmental data sets. This also serves to leverage industry work. Generally, data on the distribution of plant and animal species are hard to find. Seldom are data in digital form, but they exist as records or maps.

The USFWS maintains information about plant and animal distributions and habitats. The U.S. Geological Survey (USGS) Gap Analysis Program (GAP) efforts have also created numerous useful GIS databases on a state-by-state basis, or on a regional basis. Within the GAP, for each plant or animal of interest, a GIS file or layer is composed of distribution and habitat information. These data are available, and can be incorporated into a GIS system, along with other data sets, as part of the data inputs. It is important to note that management of diverse data types, such as GAP data, can be accomplished using a GIS and will help to leverage on-site work by avoiding the cost of generating these data independently.

Quantifying Risks

Mallard ducks, black bears, and wetland plants were selected for the GIS-based habitat vulnerability modeling in this case study because (a) sufficient habitat suitability literature was available for all taxa; (b) GIS data coverages were available and sufficient to represent published habitat requirements for all taxa; (c) the selected taxa require either wetland or shoreline conditions during at least a portion of the year; (d) the selected taxa have either undergone a population decline at some time in the study region, are recovering, or are presently listed as endangered; and (e) the selected taxa are of special interest to local, regional, or national natural resource professionals.

TABLE 6.6

Data Classes Used to Produce Habitat Suitability GIS Models for Mallard Ducks in Lower White River Region and Mississippi Alluvial Valley Ecoregion

Habitat Suitability Class in GIS Models	Lower White River Region Models	Mississippi Alluvial Valley Ecoregion Models
Wetland with fluctuating hydroperiod; trees/shrubs present; oak included; overcup oak excluded	√	
Wetland with fluctuating hydroperiod; trees/shrubs present; oak included; overcup oak included	√	
Wetland with fluctuating hydroperiod; trees/shrubs present; oak excluded	√	
Wetland with fluctuating hydroperiod; solely herbaceous plants present	√	
Wetland with fluctuating hydroperiod; no plants present	√	
Wetland with infrequent flooding or a lake, impoundment, river, or stream; trees/shrubs present; oak included; overcup oak excluded	√	
Wetland with infrequent flooding or a lake, impoundment, river, or stream; trees/shrubs present; oak included; overcup oak included	√	
Wetland with infrequent flooding or a lake, impoundment, river, or stream; trees/shrubs present; oak excluded	√	
Wetland with infrequent flooding or a lake, impoundment, river, or stream; solely herbaceous plants present	√	
Wetland with infrequent flooding or a lake, impoundment, river, or stream; no plants present	√	
Upland; agriculture	√	√
Upland; nonagriculture	√	√
Wetland; not open water; solely herbaceous plants present		√
Wetland; not open water; woody plants present		√
Open water		√

Habitat Suitability versus Risk

Habitat suitability maps in this case study depict the different parameters for each of the selected taxa, in accordance with available literature for mallard ducks (Table 6.6), black bears (Table 6.7), and wetland plants (Table 6.8). Habitat *suitability* maps are color coded to best distinguish between habitat

TABLE 6.7

Data Classes Used to Produce Habitat Suitability GIS Models for Black Bears in Lower White River Region and Mississippi Alluvial Valley Ecoregion

Habitat Suitability Classes used in GIS Models	Lower White River Region Models	Mississippi Alluvial Valley Ecoregion Models
Wetland; woody plants present; 4 tree species present; forest present in the 1950s	√	
Wetland; woody plants present; 3 tree species present; forest present in the 1950s	√	
Wetland; woody plants present; 2 tree species present; forest present in the 1950s	√	
Wetland; woody plants present; 1 tree species present; forest present in the 1950s	√	
Wetland; woody plants present; 4 tree species present; forest present solely after 1950s	√	
Wetland; woody plants present; 3 tree species present; forest present solely after 1950s	√	
Wetland; woody plants present; 2 tree species present; forest present solely after 1950s	√	
Wetland; woody plants present; 1 tree species present; forest present solely after 1950s	√	
Nonwoody wetland; 4 herbaceous plant species present; forest absent in 1950s	√	
Nonwoody wetland; 3 herbaceous plant species present; forest absent in 1950s	√	
Nonwoody wetland; 2 herbaceous plant species present; forest absent in 1950s	√	
Nonwoody wetland; 1 herbaceous plant species present; forest absent in 1950s	√	
Nonwoody wetland; no herbaceous plant species present; forest absent in 1950s	√	
Nonwoody wetland; 4 herbaceous plant species present; forest present in 1950s	√	
Nonwoody wetland; 3 herbaceous plant species present; forest present in 1950s	√	
Nonwoody wetland; 2 herbaceous plant species present; forest present in 1950s	√	
Nonwoody wetland; 1 herbaceous plant species present; forest present in 1950s	√	

(continued)

TABLE 6.7 (CONTINUED)

Data Classes Used to Produce Habitat Suitability GIS Models for Black Bears in
Lower White River Region and Mississippi Alluvial Valley Ecoregion

Habitat Suitability Classes used in GIS Models	Lower White River Region Models	Mississippi Alluvial Valley Ecoregion Models
Nonwoody wetland; no herbaceous plant species present; forest present in 1950s	√	
Open water	√	√
Wetland; trees present, >250 m from trees		√
Wetland; nonagriculture herbaceous plants present; <250 m from trees		√
Wetland; nonagriculture herbaceous plants present; >250 m from trees		√
Upland; trees present; <250 m from trees		√
Upland; nonagriculture herbaceous plants present; <250 m from trees		√
Upland; nonagriculture herbaceous plants present; >250 m from trees		√
Upland, agriculture, urban, or nonvegetated; <250 m from trees		√
Upland, agriculture, urban, or nonvegetated; >250 m from trees		√

types and patches, and are not intended to imply relative importance of habitat type(s) for any taxon. Habitat *vulnerability* maps in this case study are consistently color coded from greatest vulnerability to least vulnerability, based on standard deviations from the mean of a particular vulnerability parameter, and are intended to imply the relative risk of loss or damage to each individual patch of habitat for the taxon or guild indicated on a map.

Data and Statistics

Habitat suitability maps depict the different habitat requirements for mallard ducks (Table 6.6), black bears (Table 6.7), and wetland plants (Table 6.8). Habitat suitability maps are color coded to best distinguish between habitat types and patches. Habitat *vulnerability* maps depict the relative risk of habitat loss or damage for each taxon or guild (i.e., wetland plants) as a result of patch size, patch shape, human-induced disturbance(s), or a combination of these parameters. Habitat vulnerability maps are color coded from greatest vulnerability to least vulnerability based on standard deviations from the mean of each vulnerability parameter.

Because one or both of the assumptions of parametric statistics tests (normality and equality of variance) are violated in all of the data (which

TABLE 6.8

Data Classes Used to Produce Habitat Suitability GIS Models for Wetland Plants in the Lower White River Region and Mississippi Alluvial Valley Ecoregion

Habitat Suitability Classes Used in GIS Models
Wetland; lacustrine; littoral; rooted vascular plants present; semipermanently flooded
Wetland; lacustrine; littoral; rooted vascular plants present; permanently flooded
Wetland; lacustrine; littoral; unconsolidated bottom; permanently flooded
Wetland; palustrine; aquatic bed; rooted vascular plants present; permanently flooded
Wetland; palustrine; persistent emergent plants present; temporarily flooded
Wetland; palustrine; persistent emergent plants present; seasonally flooded
Wetland; palustrine; persistent emergent plants present; semipermanently flooded
Wetland; palustrine; broad-leaved deciduous forest present; seasonally flooded
Wetland; palustrine; broad-leaved deciduous forest present; temporarily flooded
Wetland; palustrine; broad-leaved deciduous forest present; semipermanently flooded
Wetland; palustrine; deciduous forest present; temporarily flooded
Wetland; palustrine; deciduous forest present; seasonally flooded
Wetland; palustrine; needle-leaved deciduous forest present; seasonally flooded
Wetland; palustrine; needle-leaved deciduous forest present; semipermanently flooded
Wetland; palustrine; broad-leaved deciduous shrubs present; seasonally flooded
Wetland; palustrine; broad-leaved deciduous shrubs present; temporarily flooded
Wetland; palustrine; broad-leaved deciduous shrubs present; semipermanently flooded
Wetland; palustrine; unconsolidated bottom
Wetland; palustrine; unconsolidated shore; seasonally flooded
Wetland; riverine; lower perennial; unconsolidated bottom; permanently flooded
Wetland; riverine; lower perennial; unconsolidated bottom; temporarily flooded
Wetland; riverine; lower perennial; unconsolidated bottom; seasonally flooded

is typical of ecological data), correlation analyses for all parameters were completed with Spearman rank correlation (Zar 1984), $\alpha = 0.05$. All statistical analyses were computed with Statview software (SAS Institute, v. 5.0.1, Cary, North Carolina). All GIS calculations, gradient analyses, and mappings were performed using ArcView GIS software (Environmental Systems Research Institute, v. 3.2, Redlands, California), and ATtILA GIS software (USEPA v. 2.9). To allow for visual comparison of relative habitat vulnerability across the study area landscapes, each metric is displayed as a mean value and standard deviations from the mean.

Habitat Suitability Assessment

The GIS-based habitat suitability models all use specific habitat requirements for mallard ducks, black bears, and wetland plants to the extent that digital data are available to model their ecology in prior field research. The ecological overviews of the organisms are all condensed to include the information relevant to the GIS models contained within this case study. For example, to examine mallard ducks and black bear habitat, the complexity of the landscape and the concomitant information collected across vast areas can be managed with technological and scientific approaches, while incorporating the important details of each organism's life history requirements and other information, simultaneously.

Development of Mallard Duck Winter Habitat Metrics

The Lower Mississippi Valley is the primary wintering habitat for mallard ducks in the Mississippi Flyway (Bellrose 1976), resulting in a residence of an estimated 1.6 million ducks during the winter months (Bartonek et al. 1984). A wide variety of wetland hydrologic and vegetational conditions are required to meet different habitat requirements of mallard sexes, ages, and behavioral segments of the mallard population. Generally, winter habitat conditions in the Lower Mississippi region influence all aspects of the sociobiology of mallards, which in turn affects the fecundity and survival of mallards. For example, mallard ducks in the Lower Mississippi region typically move 1.6 km to 8 km from roost sites to foraging areas. Flights of greater than 8 km are typically a response to changes in hydrology, temperature, depleted food resources, or other disturbances (Jorde et al. 1983).

Mallard ducks are omnivores that feed on available foods that include aquatic invertebrates; wetland plant seeds, fruits, rootlets, and tubers; mast from trees; agricultural grains; mollusks; insects; small fish; fish eggs; and amphibians (Heitmeyer 1985; Allen 1987). Consequently, wetlands and other open water areas are extremely important for mallard ducks because such areas are where most of the naturally occurring duck food occurs. The timing of flooding, flood depth, and the duration of wetland flooding is a critical factor in determining the diversity and availability of organisms upon which mallard ducks feed (Fredrickson 1979; Heitmeyer and Fredrickson 1981; Nichols et al. 1983; Heitmeyer 1985; Allen 1987), and much larger numbers of mallards have been observed in the Lower Mississippi Valley when these wetter conditions exist (Nichols et al. 1983), particularly in the forested wetlands (Heitmeyer 1985). Mallards typically feed from the water's surface to a maximum of 50 cm in depth (Heitmeyer and Fredrickson 1981; Krapu 1981; Batema et al. 1985; Allen 1987). Thus, the depth of flooding, the duration of flooding, and the type of vegetation, as it relates to food resources, is cited as the primary determiner of mallard duck habitat suitability (Allen 1987). Accordingly, these factors were used to establish suitable mallard habitat and to distinguish between mallard habitat patches of differing suitability characteristics.

Although some of the agricultural land in the LWRR is managed to provide winter mallard forage, agricultural grains are not a complete substitute for natural foods (Frederickson and Taylor 1982; Baldassarre et al. 1983; Jorde et al. 1984). Therefore agricultural land was not considered to be suitable mallard habitat relative to nonagricultural wetland and was not used in the habitat vulnerability assessments in this case study.

A hierarchy of mallard duck habitat suitability requirement analogues in available GIS data was used to develop habitat suitability models, which were then input to habitat vulnerability models using a number of vulnerability parameters. Black bear GIS models and wetland plant GIS models follow processes that are similar to the mallard duck GIS models, but use taxon-specific suitability parameters and GIS data.

Because all of the naturally occurring and diverse habitat requirements of mallard ducks occur in wetlands, mallards require access to a variety of wetland types, including emergent wetlands (Figure 6.13); scrub-shrub wetlands (Figure 6.14); forested wetlands (Figure 6.15); unconsolidated bottom wetlands (Figure 6.16); aquatic bed wetlands (Figure 6.17); and the open water areas of lakes, impoundments, or rivers (Figure 6.18). The wetland types described were based on the wetland classes and hydrologic modifiers used in the National Wetland Inventory (after Cowardin et al. 1979), as they apply to mallard duck habitat requirements. Specifically, mallard duck winter habitat suitability models in the LWRR and MAVE are based on optimal foraging-related habitat requirements of (a) available wetland habitat; (b) the hydroperiod within a wetland (i.e., either temporarily flooded/seasonally flooded, permanently flooded, or dry); (c) the presence of woody vegetation; and (d) the presence of desirable mast-producing oaks, specifically excluding overcup oak (*Quercus lyrata*) that produces acorns up to 2.5 cm long and are thus less suitable oak mast for forage (Allen 1987; Table 6.6).

We chose mallard ducks as one of the modeled taxa because they are abundant and ubiquitous, are dependent upon wetlands for most of the year, have well-documented habitat requirements, require habitat that can be readily

FIGURE 6.13
An emergent wetland in the Lower White River Region.

FIGURE 6.14
A scrub-shrub wetland across a pool with duckweed (*Lemna* sp.) and watermeal (*Wolffia* sp.) on the surface in Lower White River Region.

FIGURE 6.15
A forested wetland in the Lower White River Region.

FIGURE 6.16
Unconsolidated bottom wetland types in the Lower White River Region may be (a) emergent or (b) forested on their perimeters.

mapped using GIS data, and are a species that has recovered from previously lower numbers in the study area and throughout the Mississippi Flyway.

Mallard ducks are a key wetland species, and represent the multitude of waterfowl that forage, rest, and breed in wetlands. Their cosmopolitan presence in wetlands of the continent and ubiquity among many areas speak to their importance as indicators of wetland condition and thus importance in wetland landscape assessments. Additional attention has also been focused on terrestrial megafauna and plants that are dependent upon wetland and associated aquatic ecosystems and are addressed in the next sections.

Development of Black Bear Habitat Metrics

Black bears are an important species for wetland habitat metrics because they are a megafaunal species that is understood to be recovering from extirpation, and they are a cosmopolitan species that makes use of both upland and wetland environments throughout their home range. They are found throughout North America and, in the LWRR and MAVE, have a mean female

FIGURE 6.17
An aquatic bed with submersed and emergent vegetation present.

home range from 9 to 12 km². Male home ranges have been reported as large as 116 to 148 km² (Pelton 1982; Klepinger and Norton 1983; Smith 1985), while others have reported male home ranges from 13 to 24 km² (Rogers 1992). The home ranges of male and female black bears are primarily dependent on the availability of resources (Jonkel and Cowan 1971; Amstrup and Beecham 1976; LeCount 1980; Reynolds and Beecham 1980; McArthur 1981; Elowe 1984; Rogers 1987), and are also influenced by population density, age of the

FIGURE 6.18
A portion of the White River in the Lower White River Region study area. River areas provide areas for animals to feed, drink, and rest.

individual, and seasonal conditions (Pelton 1982). In 1997, 187 black bears were legally hunted solely in the mountainous regions in Arkansas using muzzleloader, modern gun, archery, and crossbow hunting techniques. As a result of increasing numbers of black bears in the vicinity of the LWRR, hunting of black bears within selected areas to the west of the Mississippi River, using modern guns and other hunting techniques, became legal again in 1999 excluding the White River National Wildlife Refuge (Arkansas Game and Fish Commission 1998). Today, black bear numbers are on the increase, marked by a reestablishment of licensed hunting in the past decade.

Black bears are omnivores that typically feed on easily digestible vegetative foods that are high in nutrients and low in cellulose (Rogers 1976, 1987; Herrero 1979). Thus the typical black bear diet consists of fruits, nuts, acorns, insects, and early-sprouting green vegetation (Mealey 1975; Herrero 1979; Rogers 1987). When naturally growing food items are scarce, black bears may alternatively feed on agricultural crops, such as orchard fruits or corn, or at human-constructed food sources, such as centralized refuse disposal sites or the residences of humans (Harger 1967; Bray 1974; Rogers 1976; Hugie 1979; Landers et al. 1979; Beeman and Pelton 1980; Rogers 1987). Scarcity of naturally growing foods for black bears has been positively correlated with occurrences of bear cannibalism (Tietje et al. 1986) and the occurrences of bear–human interactions.

Black bears in the LWRR and MAVE may remain active throughout the winter to feed on corn and other foods if other naturally growing foods are scarce (Carpenter 1973; Matula 1974; Lindzey et al. 1976; Rogers 1976; Hamilton 1978; Hamilton and Marchington 1980; Elowe 1984). Black bear predation on vertebrates is relatively rare but such captures in the LWRR and MAVE may include newborn deer, nestling birds, fish, or other animals whose escape is hampered (Rowan 1928; Barmore and Stradley 1971; Frame 1974; Cardoza 1976; Ozoga and Verme 1984).

Wetlands and other open water areas are extremely important for black bears to survive because, aside from providing much of the food resources that they require, such areas are frequently the sole resource for drinking water and for providing water in which they can cool themselves. The home ranges of black bears are also closely tied to forested areas (Herrero 1979; Hugie 1979; Pelton 1982), and are limited by the fact that much of the remaining forested areas in the LWRR and MAVE are fragmented and therefore relatively inaccessible (Cowan 1972; Maehr and Brady 1984; Twedt et al. 1999). Black bears in relatively fragmented landscapes, like the MAVE, are frequently observed in forest openings and clearings, which may provide a relatively higher degree of edge vegetation diversity than core forest areas (Herrero 1979; Hugie 1979).

Wetlands, particularly in riparian areas, are used by black bears for seasonal foraging, denning, cover for *escape*, and as travel corridors. Thus, as a result of the diverse resource and travel corridor requirements, black bears require access to a variety of wetland types and resources, including emergent wetlands; scrub-shrub wetlands; forested wetlands; unconsolidated bottom wetlands; aquatic bed wetlands; and the open water areas of lakes, impoundments, or rivers. Specifically, the black bear habitat suitability models in the LWRR and MAVE are based on optimal habitat requirements of (a) available wetland habitat, (b) the presence of woody vegetation, (c) the number of plant species within a patch, and (d) evidence of forest disturbance since the 1950s (Table 6.7). We chose black bears as one of the modeled taxa for this case study because they are a recovering species in the region, are dependent upon wetlands for most of the year, have well-documented habitat requirements, and require habitat that can be readily mapped using GIS data.

Development of Wetland Plant Habitat Metrics

Because the previously mentioned animals are dependent on wetlands (specifically, wetland plants), it is important to investigate wetland plant metrics and the associated health of wetlands. Wetland plants are adapted to surviving in soils that are saturated with water (and consequently have low oxygen content). Plants that flourish in wetlands are thus generally referred to as hydrophytes, but may also include plants that can survive with only brief periods of soil saturation, and the low-oxygen soil conditions that accompany soil saturation (Reed 1988; Lyon and Lyon 2011). Wetland plant species in the LWRR and the MAVE are numerous, and provide cover and forage for other organisms in emergent wetlands; scrub-shrub wetlands; forested wetlands; aquatic bed wetlands; and in the littoral zones of some lakes, impoundments, and rivers. Accordingly, wetland plant habitat suitability models in the LWRR and MAVE are based on the optimal wetland plant habitat requirements of (a) the presence of wetland conditions (particularly, the presence of wetland soil types; Mitsch and Gosselink 1993) and (b) the hydroperiod within each wetland type (Table 6.8). We chose the collective group (i.e., a guild) of regional wetland plant species for this case study because they are strictly limited to the geographic extent of wetlands, are strictly dependent upon wetland hydrology, have well-documented habitat requirements, have well-understood physiological responses to hydrologic and other physical disturbances, and require habitat that can be readily mapped using GIS data.

The interactivity of organisms in wetlands allows us to observe and use the data collected from waterfowl, mammal, and plant data for a combined look at availability for use (i.e., habitat suitability) and risk of loss or alteration impacts (i.e., vulnerability).

Human-Related Features and Wetland Mitigation

When assessing the biological elements of wetlands, the human component must be evaluated as well. This work has demonstrated that there are a number of features of wetlands and their surrounding landscapes that may be influenced or related to humans. Regulatory agencies and the public now require additional examination of human-related issues, and the presence or

FIGURE 6.19
A low-altitude image demonstrated the capability to view drainage ways via the dark tone of moisture-laden soils created by the light quenching effect of moisture. The presence of trees is given by the crown shape, dark tones, and vigorous growing plants. The dark toned areas on the boundary of the forested area may have been former wetland areas, now cropped, which could lead to the development of wetland enhancements (i.e., restoration or creation) in the future, depending upon decisions of partners in the region.

absence of these features can be important in selecting lands for construction of wetlands for habitat enhancements or mitigation (Figure 6.19).

Restricted properties may also influence the selection of land appropriate for use as a constructed wetland. These conditions are known to industry, and available records or databases can be consulted. Sources can include state and local historical societies, and anthropological and archaeological groups. The data derived from records and maps can be incorporated in the GIS to be used in selection.

Incompatible land ownership or land not available for easements or for sale present a real problem (Figure 6.20) for privately and sometimes publically owned property. Prevailing conditions can be noted and stored for analyses. However, schools, cemeteries, restricted federal and state ownership, and other ownership are factored. Future considerations may include criterion such as house counts in a given area or population.

All the relevant human-related factors must be assessed to provide information about risk, such as in terms of wetland habitat functions or water quality functions. These factors are essential for providing decision makers with the optimal decision-making tools. Whether there are interests that involve wildlife, conservation, sustainable development, or other policy-related and planning exercises, the need for risk analysis is a key component of wetland landscape assessments.

FIGURE 6.20
Higher altitude imagery shows many of the same types of features as in Figure 6.19. The drainage ways stand out from the surrounding soils and crops by their relative dark tone and the sinuous pathway downhill.

Characterizing Habitat Vulnerability

Habitat vulnerability is essentially a risk-based approach, and in the case of the LWRR and MAVE is associated with human activities, such as urban-built environments, road development, and agricultural activity and (often associated) patch configuration such as size or shape. Additional attention to these human-related activities can enable the wetland practitioner to make the important socioecological connections that are critical to successful wetland landscape assessment.

Risks to all of the stakeholders and habitats are always a factor. In this case study, habitat vulnerability was determined for all suitable wetland habitat patches in the LWRR and MAVE for mallard duck, black bear, and wetland plants if a habitat patch was greater than 2 ha.

Habitat vulnerability measures (Table 6.1) within habitat patches for each taxon are based on patch size, patch shape, human-induced disturbances, and combinations of these metrics. All of the habitat vulnerability metrics within a habitat patch may affect the likelihood that a particular habitat patch will rebound after patch disturbance(s). That is, habitat vulnerability metrics were based on predicted habitat degradation as a result of patch destruction (i.e., total loss), patch fragmentation (i.e., partial loss), or patch degradation (i.e., stress) (after Odum 1985).

Patch size metrics (Table 6.1) were based on previously observed ecosystem trends regarding the effects of patch size on habitat quality for specific taxa in many different regions (e.g., MacArthur and Wilson 1967; Simberloff and

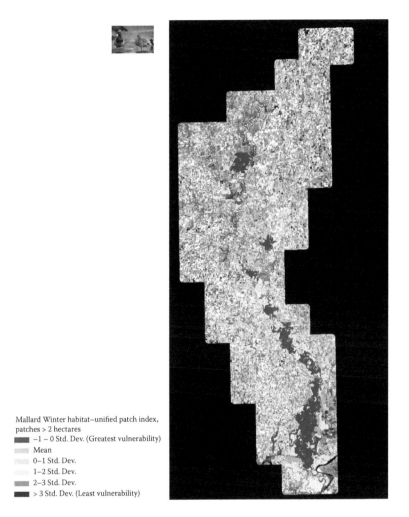

FIGURE 6.21 (a)
(See color insert.) Lower White River Region (LWRR) (a) mallard duck, (b) black bear, and (c) wetland plant habitat vulnerability models (>2 ha) in terms of habitat patch size and shape.

Wilson 1970; Diamond 1974; Forman et al. 1976; Pickett and Thompson 1978; Soule et al. 1979; Hermy and Stieperaere 1981; Van der Valk 1981; Simberloff and Abele 1982; McDonnell and Stiles 1983; Harris 1984; McDonnell 1984; Moller and Rordam 1985; Brown and Dinsmore 1986; Dzwonko and Loster 1988; Gutzwiller and Anderson 1992; Opdam et al. 1993; Hamazaki 1996; Kellman 1996; Bastin and Thomas 1999; McIntyre and Wiens 1999a, 1999b; Twedt and Loesch 1999; Jones et al. 2000; Lopez et al. 2002; Lopez and Fennessy 2002).

Accordingly, the *patch size* habitat vulnerability models map the habitat patch area, habitat patch perimeter length, and habitat patch interior-to-edge ratio in the LWRR (Figure 6.21) and the MAVE (Figure 6.22). Smaller habitat patches

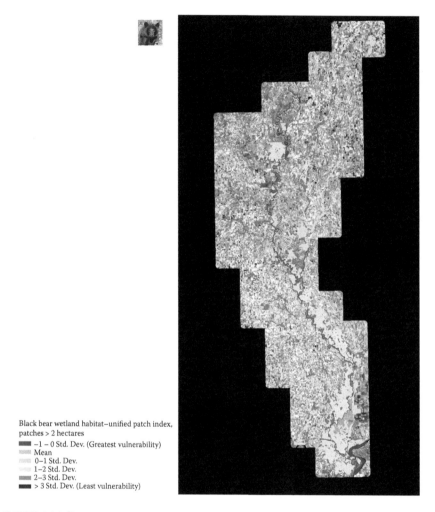

Black bear wetland habitat–unified patch index,
patches > 2 hectares
■■■ −1 – 0 Std. Dev. (Greatest vulnerability)
▨▨▨ Mean
▨▨▨ 0–1 Std. Dev.
▨▨▨ 1–2 Std. Dev.
■■■ 2–3 Std. Dev.
■■■ > 3 Std. Dev. (Least vulnerability)

FIGURE 6.21 (b)
(See color insert.) Lower White River Region (LWRR) (a) mallard duck, (b) black bear, and (c) wetland plant habitat vulnerability models (>2 ha) in terms of habitat patch size and shape.

(as measured by area, perimeter, or interior-to-edge ratio) are relatively less likely to rebound from disturbances (i.e., are more likely to be fragmented or destroyed after changes in hydrology, destruction of vegetation, or the establishment of opportunistic flora and fauna) than larger habitat patches.

The ecological vulnerability metrics regarding patch shape (Table 6.1) were based on trends previously observed for specific taxa, in many different regions (e.g., MacArthur and Wilson 1967; Gilpin 1981; McDonnell and Stiles 1983; Harris 1984; McDonnell 1984; Gutzwiller and Anderson 1992; Hamazaki 1996; Kellman 1996; Bastin and Thomas 1999; Jones et al. 2000; Lopez et al.

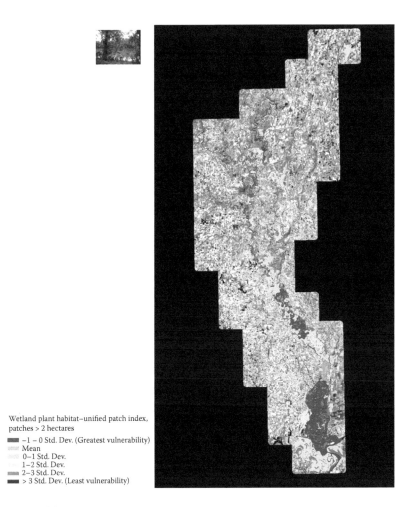

Wetland plant habitat–unified patch index,
patches > 2 hectares

▬ −1 – 0 Std. Dev. (Greatest vulnerability)
▬ Mean
▬ 0–1 Std. Dev.
▬ 1–2 Std. Dev.
▬ 2–3 Std. Dev.
▬ > 3 Std. Dev. (Least vulnerability)

FIGURE 6.21 (c)
(See color insert.) Lower White River Region (LWRR) (a) mallard duck, (b) black bear, and (c) wetland plant habitat vulnerability models (>2 ha) in terms of habitat patch size and shape.

2002; Lopez and Fennessy 2002), and include patch *sinuosity* (after Bosch 1978; Davis 1986) in the LWRR (Figure 6.21) and the MAVE (Figure 6.22).

Patches with a smaller sinuosity index are less winding or convoluted in shape, thus are relatively less likely to rebound from disturbances than patches with a greater sinuosity index. That is, smaller index values indicated that a habitat patch has a greater likelihood of fragmentation or loss as a result of environmental change, such as changes in hydrology, destruction of vegetation, or the establishment of opportunistic flora and fauna. Another metric of patch shape is patch *circularity* (after Stoddart 1965; Unwin 1981) in the LWRR (Figure 6.21) and the MAVE (Figure 6.22). Habitat patches with a smaller circularity index are more circular in shape, thus

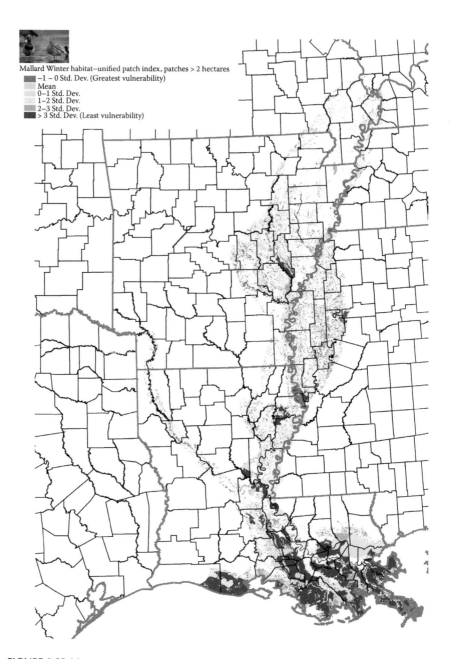

FIGURE 6.22 (a)
(See color insert.) Mississippi Alluvial Valley Ecoregion (MAVE) (a) mallard duck, (b) black bear, and (c) wetland plant habitat vulnerability models (>2 ha) in terms of habitat patch size and shape.

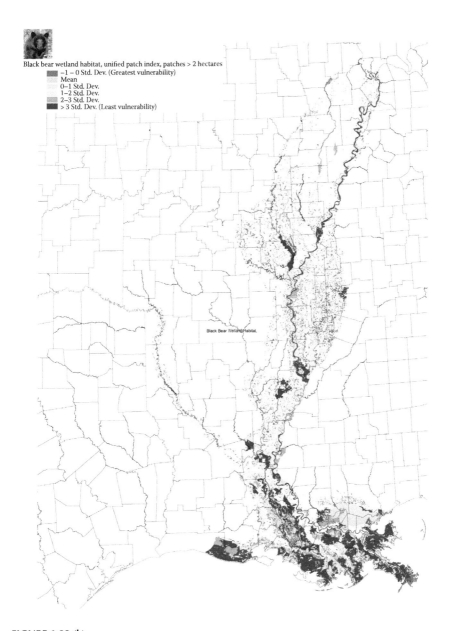

Black bear wetland habitat, unified patch index, patches > 2 hectares
■ −1 − 0 Std. Dev. (Greatest vulnerability)
Mean
0–1 Std. Dev.
1–2 Std. Dev.
2–3 Std. Dev.
■ > 3 Std. Dev. (Least vulnerability)

FIGURE 6.22 (b)
(See color insert.) Mississippi Alluvial Valley Ecoregion (MAVE) (a) mallard duck, (b) black bear, and (c) wetland plant habitat vulnerability models (>2 ha) in terms of habitat patch size and shape.

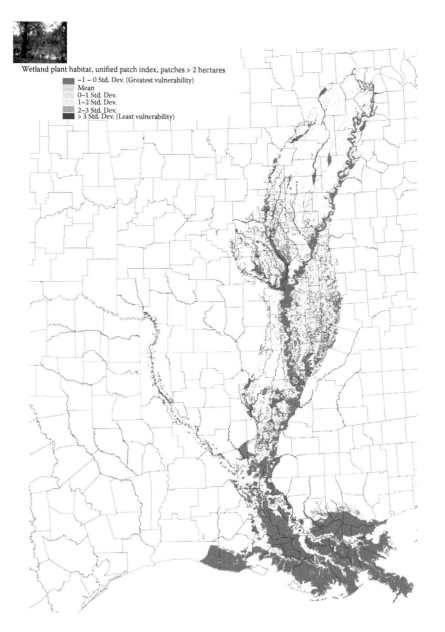

FIGURE 6.22 (c)
(See color insert.) Mississippi Alluvial Valley Ecoregion (MAVE) (a) mallard duck, (b) black bear, and (c) wetland plant habitat vulnerability models (>2 ha) in terms of habitat patch size and shape.

are relatively less likely to rebound from disturbances than patches with a larger circularity index. That is, smaller index values indicate that a habitat patch has a greater likelihood of fragmentation or loss as a result of environmental change, such as changes in hydrology, destruction of vegetation, or the establishment of opportunistic flora and fauna. The unified patch index (Table 6.1) multiplicatively combines the patch interior-to-edge metric, the patch sinuosity index, and the patch circularity index into a single metric that depicts patch area, patch perimeter, and patch shape simultaneously as a measure of habitat patch vulnerability in the LWRR (Figure 6.21) and the MAVE (Figure 6.22).

Human-induced disturbance factors within habitat patches in the LWRR and MAVE (Table 6.1) are based on previously observed positive correlations between ecosystem degradation and amount of land cover conversion during road construction, road maintenance, and other human activities (e.g., Connell and Slatyer 1977; Van der Valk 1981; Ehrenfeld 1983; Johnston 1989, 1994; Scott et al. 1993; Poiani and Dixon 1995; Strittholt and Boerner 1995; Jenning 1995; Wilcox 1995; Ogutu 1996; Stiling 1996; Heggem et al. 2000; Lopez et al. 2002; Lopez and Fennessy 2002). Thus the directly measurable human-induced disturbance metrics within habitat patches are road length and road density in the LWRR (Figure 6.23) and the MAVE (Figure 6.24).

Patches with greater total road length or road density (or their combined index value; Table 6.1) are relatively less likely to rebound from disturbances than patches with a lesser road presence. That is, greater road length, road density, or road index values indicate that a patch has a greater likelihood of fragmentation or loss as a result of environmental change, such as changes in hydrology, destruction of vegetation, or the establishment of opportunistic flora and fauna. The increased presence of roads within a patch may also bring about a decrease in patch size or shape metrics. The road index additively combines road length and road density metrics to depict the combined potential habitat degradation effects of road construction and maintenance on habitat patches in the LWRR (Figure 6.23) and the MAVE (Figure 6.24).

The indirect metrics of human-induced disturbance within habitat patches were human population density in 1990, estimated human population density in 2011 (i.e., future human population density), and estimated human population density change from 1990 to 2011 for the LWRR (Figure 6.23) and the MAVE (Figure 6.24). The human population density change metric was normalized to a positive number by adding 50 to the calculation for the LWRR, and by adding 8,000 to the calculation for the MAVE models (Table 6.1). Habitat patches that existed within census block groups (Table 6.3 and Table 6.4) with a greater population density now or in the future were relatively less likely to rebound from disturbances than patches that existed in areas of lesser population density now or in the future. That is, greater human population density values indicated that a habitat patch has a greater

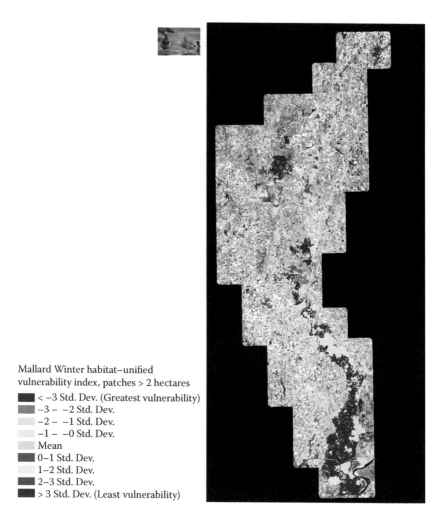

Mallard Winter habitat–unified
vulnerability index, patches > 2 hectares
■ < −3 Std. Dev. (Greatest vulnerability)
■ −3 – −2 Std. Dev.
  −2 – −1 Std. Dev.
  −1 – −0 Std. Dev.
  Mean
■ 0–1 Std. Dev.
  1–2 Std. Dev.
■ 2–3 Std. Dev.
■ > 3 Std. Dev. (Least vulnerability)

FIGURE 6.23 (a)
(See color insert.) Lower White River Region (LWRR) (a) mallard duck, (b) black bear, and (c) wetland plant habitat vulnerability models (>2 ha) in terms of human-induced disturbance factors.

likelihood of fragmentation or loss as a result of environmental change, such as changes in hydrology, destruction of vegetation, or the establishment of opportunistic flora and fauna.

The increased presence of humans near habitat patches may also bring about a decrease in patch size metrics, patch shape metrics, an increase in road length, or an increase in road density. The unified human index additively combines road length, road density, and human population density change (from 1990 to 2011) to depict the combined effects of direct and indirect human-induced disturbance in the LWRR (Figure 6.23) and the MAVE

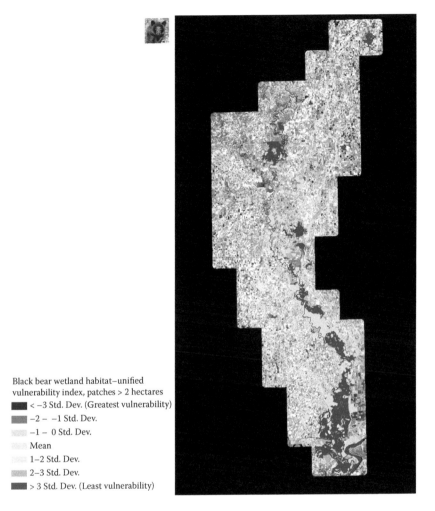

Black bear wetland habitat–unified
vulnerability index, patches > 2 hectares
■ < −3 Std. Dev. (Greatest vulnerability)
■ −2 – −1 Std. Dev.
  −1 – 0 Std. Dev.
  Mean
  1–2 Std. Dev.
  2–3 Std. Dev.
■ > 3 Std. Dev. (Least vulnerability)

FIGURE 6.23 (b)
(See color insert.) Lower White River Region (LWRR) (a) mallard duck, (b) black bear, and (c)
wetland plant habitat vulnerability models (>2 ha) in terms of human-induced disturbance
factors.

(Figure 6.24). The unified vulnerability index (Table 6.1) additively combines
the unified patch index, direct human-induced disturbance metrics, and
indirect human-induced disturbance metrics, so that the combined effects of
patch size, patch shape, road metrics, and human population density metrics
can be depicted simultaneously in the LWRR (Figure 6.23) and the MAVE
(Figure 6.24).

Because the GIS models are based on 30 m resolution data (e.g., Arkansas
Gap Analysis Project [AR-GAP] in the LWRR; National Land Cover Data

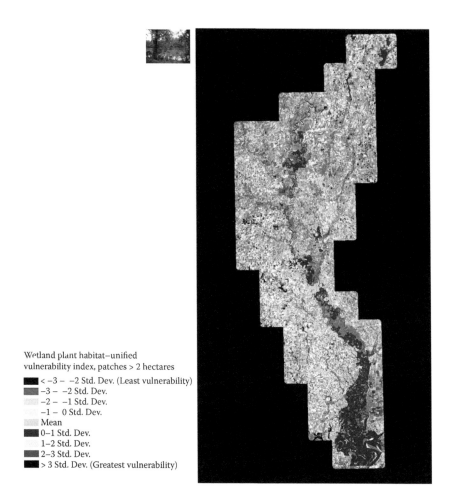

Wetland plant habitat–unified
vulnerability index, patches > 2 hectares

■ < −3 − −2 Std. Dev. (Least vulnerability)
■ −3 − −2 Std. Dev.
  −2 − −1 Std. Dev.
  −1 − 0 Std. Dev.
  Mean
■ 0–1 Std. Dev.
  1–2 Std. Dev.
■ 2–3 Std. Dev.
■ > 3 Std. Dev. (Greatest vulnerability)

FIGURE 6.23 (c)
(See color insert.) Lower White River Region (LWRR) (a) mallard duck, (b) black bear, and (c) wetland plant habitat vulnerability models (>2 ha) in terms of human-induced disturbance factors.

[NLCD] in the MAVE), as with all similar analyses, caution should be exercised when using the results at fine scales.

One of the important applications of the wetland analyses includes assessing the wildlife refuges of the LWRR, with an emphasis on the relative priorities of the landscape. An approach that is often used is mapping and listing the relative condition of the resources in a specific area, such a refuge or conservation area, for decision-making purposes.

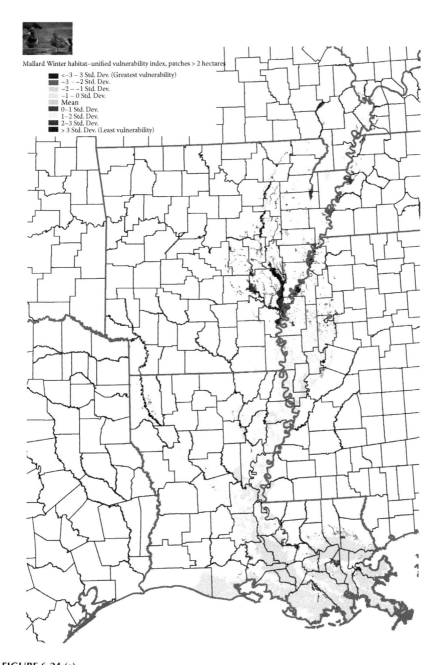

Mallard Winter habitat–unified vulnerability index, patches > 2 hectares

- ■ <−3 – 3 Std. Dev. (Greatest vulnerability)
- ■ −3 – −2 Std. Dev.
- ░ −2 – −1 Std. Dev.
- ░ −1 – 0 Std. Dev.
- ▒ Mean
- ■ 0–1 Std. Dev.
- 1–2 Std. Dev.
- 2–3 Std. Dev.
- ■ > 3 Std. Dev. (Least vulnerability)

FIGURE 6.24 (a)
(See color insert.) Mississippi Alluvial Valley Ecoregion (MAVE) (a) mallard duck, (b) black bear, and (c) wetland plant habitat vulnerability models (>2 ha) in terms of human-induced disturbance factors.

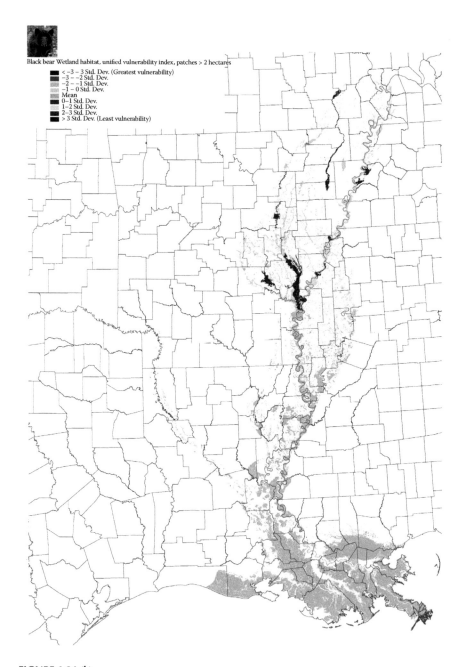

FIGURE 6.24 (b)
(See color insert.) Mississippi Alluvial Valley Ecoregion (MAVE) (a) mallard duck, (b) black bear, and (c) wetland plant habitat vulnerability models (>2 ha) in terms of human-induced disturbance factors.

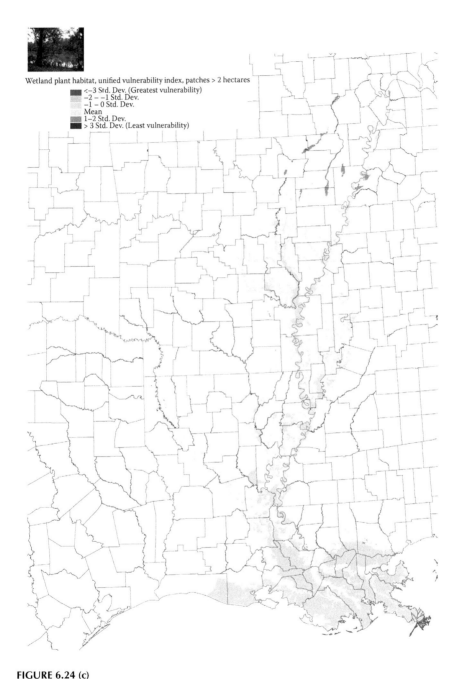

Wetland plant habitat, unified vulnerability index, patches > 2 hectares
▮ <–3 Std. Dev. (Greatest vulnerability)
▯ –2 – –1 Std. Dev.
▯ –1 – 0 Std. Dev.
▯ Mean
▮ 1–2 Std. Dev.
▮ > 3 Std. Dev. (Least vulnerability)

FIGURE 6.24 (c)
(See color insert.) Mississippi Alluvial Valley Ecoregion (MAVE) (a) mallard duck, (b) black bear, and (c) wetland plant habitat vulnerability models (>2 ha) in terms of human-induced disturbance factors.

Resource Applications of Results

Habitat Elements

Waterfowl are excellent indicators of wetland condition and create a link with cross-boundary areas throughout the American Continent. Mallard duck are ubiquitous among many types of wetlands and they are a hearty species that are relatively simple to observe, determine species counts for, and develop habitat use studies for. Although there are a plethora of wildfowl species that frequent wetlands, the amount of information about mallard habitat requirements makes for an excellent implementation for landscape models using remote sensing and GIS data.

The previously discussed mallard duck winter habitat vulnerability models were used to assess the 895 km² of federal refuge lands within the 8,921 km² LWRR. LWRR mallard duck winter habitat (Table 6.9) exists within thirteen Arkansas counties (Figure 6.25), of which a portion intersects the seventy-two separate parcels of land that comprise the White River National Wildlife Refuge (NWR), Cache River NWR, and Bald Knob NWR (Figure 6.26 and Figure 6.27). The USFWS (DeWitt, Arkansas) supplied the boundary of each national wildlife refuge zone (RZ, hereafter) as ArcView shape files (Environmental Systems Research Institute, v. 3.2, Redlands, California).

TABLE 6.9

Area of Potential Mallard Duck Winter Habitat in the Lower White River Region

Mallard Duck Winter Habitat Suitability Class	Habitat Area (ha)
Wetland with fluctuating hydroperiod; trees/shrubs present; oak included; overcup oak excluded	82,164
Wetland with fluctuating hydroperiod; trees/shrubs present; oak included; overcup oak included	21,173
Wetland with fluctuating hydroperiod; trees/shrubs present; oak excluded	77,036
Wetland with fluctuating hydroperiod; solely herbaceous plants present	8,564
Wetland with fluctuating hydroperiod; no plants present	5,730
Wetland with infrequent flooding or a lake, impoundment, river, or stream; trees/shrubs present; oak included; overcup oak excluded	1,924
Wetland with infrequent flooding or a lake, impoundment, river, or stream; trees/shrubs present; oak included; overcup oak included	486
Wetland with infrequent flooding or a lake, impoundment, river, or stream; trees/shrubs present; oak excluded	7,344
Wetland with infrequent flooding or a lake, impoundment, river, or stream; solely herbaceous plants present	2,772
Wetland with infrequent flooding or a lake, impoundment, river, or stream; no plants present	19,875
Upland; agriculture	423,763
Upland; nonagriculture	241,246
Total	892,077

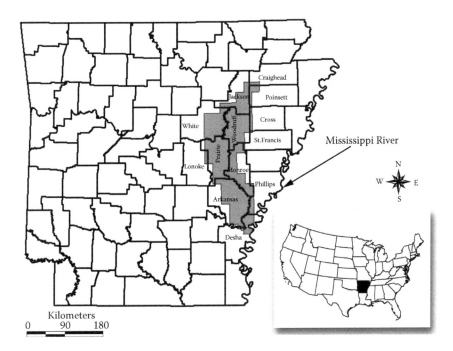

FIGURE 6.25
Orientation map of Lower White River Region (LWRR), including the thirteen counties that intersect the LWRR study area.

An important element of a wetland landscape assessment outcome is a thorough and detailed measure of the habitat elements within the study area. To examine the influence of the four largest RZs (contributing 83% of the area to all RZs in the LWRR) we isolated their calculations (i.e., RZs 72, 67, 62, and 8) from certain detailed analyses (Table 6.10), but correlations between habitat area and refuge area did not substantially change (Table 6.10). The rank order of RZs with regard to RZ area, RZ perimeter, area of habitat patches within RZ, and mean percent contribution of habitat patches within a RZ suggests that larger RZs tend to capture larger wetland habitat patches, but this relationship was nonlinear, and the relationship was most clearly demonstrated among either relatively larger RZs or relatively smaller RZs (Table 6.11). Eight specific RZs demonstrated how a larger refuge does not necessarily result in larger mallard habitat within the refuge. RZ 29 was relatively large with very small intersecting habitat patches; and RZs 48, 44, 40, 32, 18, 13, and 11 were relatively small with large intersecting habitat patches.

Table 6.10 summarizes the general trend that RZ area is weakly negatively correlated with (a) original area of habitat patches inside and outside of the RZ ($\alpha = -0.141$, $P = <0.0001$); (b) habitat patch perimeter within a RZ ($\alpha = -0.095$, $P = <0.0001$); (c) habitat patch interior-to-edge ratio within a RZ ($\alpha = -0.188$, $P = <0.0001$); (d) population density in the year 1990 ($\alpha = -0.214$, $P = <0.0001$);

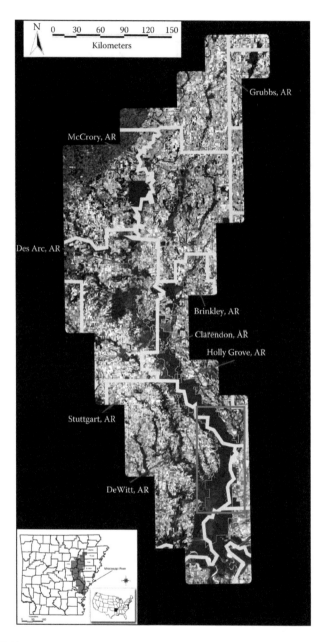

FIGURE 6.26 (a)
(See color insert.) (a) Single near-infrared band image of the Lower White River Region with selected towns indicated with blue arrows. Wildlife refuge zones are indicated by white polygons. (b) Same image displayed in a three-band false-color infrared composite image. (c) Same image displayed in a three-band false-color "enhanced vegetation" infrared composite image.

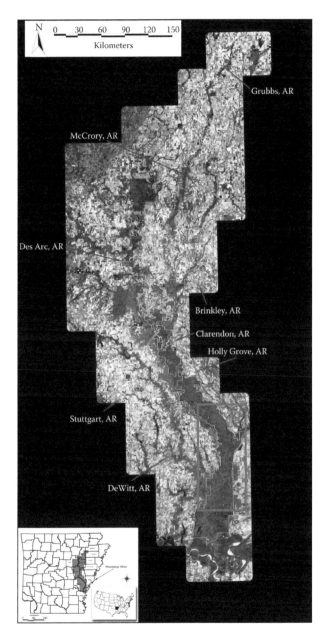

FIGURE 6.26 (b)
(See color insert.) (a) Single near-infrared band image of the Lower White River Region with selected towns indicated with blue arrows. Wildlife refuge zones are indicated by white polygons. (b) Same image displayed in a three-band false-color infrared composite image. (c) Same image displayed in a three-band false-color "enhanced vegetation" infrared composite image.

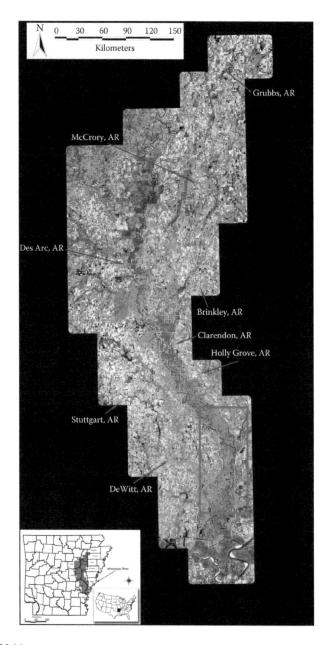

FIGURE 6.26 (c)
(See color insert.) (a) Single near-infrared band image of the Lower White River Region with selected towns indicated with blue arrows. Wildlife refuge zones are indicated by white polygons. (b) Same image displayed in a three-band false-color infrared composite image. (c) Same image displayed in a three-band false-color "enhanced vegetation" infrared composite image.

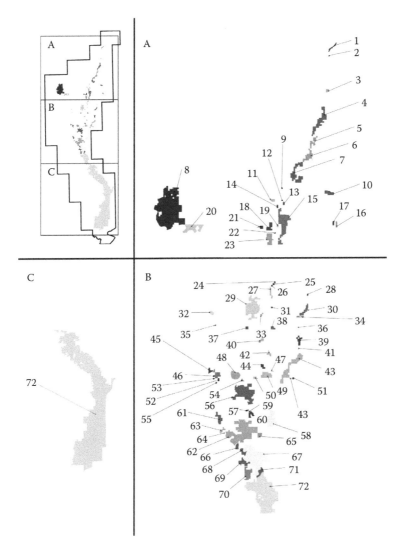

FIGURE 6.27
(See color insert.) Seventy-two federal refuge zones in the Lower White River Region were compared using mallard duck winter habitat vulnerability models.

(e) estimated population density in the year 2011 ($\alpha = -0.233$, P = <0.0001); and (f) estimated population density change from 1990 to 2011 in a RZ ($\alpha = -0.039$, P = <0.031).

To hold aside the influence of the four largest RZs we excluded them from the metric comparisons (see explanation earlier in this chapter), resulting in trends similar to the full group of seventy-two RZs (Table 6.10). Rankings of mallard duck wetland habitat patch metrics and indexes (Table 6.12) indicated that (a) RZs 72, 67, 62, and 8 were relatively more vulnerable to disturbance as

TABLE 6.10

Summary of Spearman Rank Correlation Between Federal Refuge Zone and Habitat Vulnerability Parameters

Habitat Patch Metric or Index Name	Area of All Federal Refuge Zones	Area of All Federal Refuge Zones Except Four Largest Refuge Zones
Area of habitat patches within a refuge zone	0.127****	0.116**
Percent contribution of habitat patches within a refuge zone	−0.723****	−0.503****
Original (i.e., inside and outside) area of the habitat patches for a refuge zone	−0.141****	−0.149****
Habitat patch perimeter within a refuge zone	−0.095****	−0.125***
Habitat patch interior:edge ratio within a refuge zone	−0.188****	−0.148****
Habitat patch sinuosity index within a refuge zone	ns	ns
Habitat patch circularity index within a refuge zone	ns	ns
Unified patch index within a refuge zone	ns	ns
Habitat patch road density within a refuge zone	0.032****	-0.026****
Habitat patch road length within a refuge zone	ns	ns
Habitat patch road index within a refuge zone	ns	ns
Population density in the year 1990 within a refuge zone	−0.214****	−0.113**
Estimated population density in the year 2011 within a refuge zone	−0.233****	−0.114**
Estimated population density change from 1990 to 2001 within a refuge zone	−0.039*	ns
Unified human index within a refuge zone	ns	ns
Unified vulnerability index within a refuge zone	ns	ns

Correlation (Rho) values and significance are shown; ns = not significant, * = 0.05, ** = 0.01, *** = 0.001, **** = 0.0001. All calculations are based upon habitat patches > 2 ha.

a result of a smaller habitat patch interior-to-edge ratio; (b) RZs 72, 67, and 8 were relatively more vulnerable to disturbance as a result of a smaller habitat patch area; (c) RZs 72 and 8 are relatively more vulnerable to disturbance as a result of a smaller habitat patch perimeter, sinuosity index, and circularity index; and (d) RZ 8 was relatively more vulnerable to disturbance as a result of a smaller habitat patch unified patch index (Table 6.12).

TABLE 6.11

Rank of each of the 72 Federal Refuge Zones in the Lower White River Region

	Rank of Federal Wildlife Refuge Zone Metric (Zones 1–72)			
Federal Refuge Zone Area (ha)	Federal Refuge Zone Perimeter (km)	Mean Original Habitat Patch Area within Federal Refuge Zone	Mean Percent Contribution of Each Habitat Patch to Federal Refuge Zone	Relative Metric Value Guide
12	12	12	72	Lowest
64	53	24	29	
2	2	25	8	
22	22	2	67	
53	64	53	15	
9	9	44	62	
59	26	29	4	
26	25	64	56	
25	31	40	48	
41	28	22	7	
31	24	50	43	
28	51	17	58	
24	55	9	44	
47	41	59	6	
51	47	19	18	
55	52	42	40	
57	59	33	20	
52	13	26	24	
14	14	34	61	
13	19	14	30	
19	57	18	12	
33	17	16	23	
17	54	28	69	
35	3	47	5	
54	37	48	71	
3	50	60	42	
16	33	51	50	
50	35	3	60	
37	11	54	46	
34	16	55	39	
21	38	30	10	
1	66	31	34	
11	32	71	25	
38	65	8	17	
42	21	6	70	

(continued)

TABLE 6.11 (CONTINUED)

Rank of each of the 72 Federal Refuge Zones in the Lower White River Region

	Rank of Federal Wildlife Refuge Zone Metric (Zones 1–72)			
Federal Refuge Zone Area (ha)	Federal Refuge Zone Perimeter (km)	Mean Original Habitat Patch Area within Federal Refuge Zone	Mean Percent Contribution of Each Habitat Patch to Federal Refuge Zone	Relative Metric Value Guide
45	44	61	27	
40	40	39	16	
44	34	4	65	
63	45	35	36	
66	42	45	33	
32	63	5	45	
27	60	7	19	
65	1	21	68	
60	10	27	3	
68	68	15	66	
36	71	41	21	
39	27	13	49	
71	18	69	14	
10	48	10	1	
18	49	1	54	
30	36	46	63	
5	46	65	37	
46	39	37	2	
61	61	66	9	
69	23	58	53	
6	30	23	35	
23	69	36	59	
49	70	63	26	
48	6	68	51	
70	5	43	47	
7	20	67	38	
20	58	57	32	
4	4	20	64	
58	7	56	22	
43	15	38	55	
15	43	52	13	
29	29	11	11	
56	56	72	28	

(continued)

TABLE 6.11 (CONTINUED)

Rank of each of the 72 Federal Refuge Zones in the Lower White River Region

		Rank of Federal Wildlife Refuge Zone Metric (Zones 1–72)		
Federal Refuge Zone Area (ha)	Federal Refuge Zone Perimeter (km)	Mean Original Habitat Patch Area within Federal Refuge Zone	Mean Percent Contribution of Each Habitat Patch to Federal Refuge Zone	Relative Metric Value Guide
67	67	62	31	
62	62	32	41	
8	8	70	57	
72	72	49	52	Highest

Table 6.13 describes the seventy-two RZs ranked by *human-induced disturbance* metrics and indexes, with the median indicated for each value range. The ranks of RZs indicated that (a) RZs 72, 67, 62, and 8 were relatively more vulnerable to disturbance as a result of a greater habitat patch road density, road length, road index, and unified human index; (b) RZs 67, 62, and 8 were relatively more vulnerable to disturbance as a result of a greater habitat patch population density in 1990 and expected population density in 2011; (c) RZ 8 was relatively more vulnerable to disturbance as a result of a greater habitat patch population density change expected in the year 2011; and (d) RZs 72, 67, 62, and 8 were relatively less vulnerable to disturbance as indicated by the relatively larger unified vulnerability index (Table 6.13).

For wetland habitat patches less than or equal to 2 ha there was a greater percent contribution of the original habitat patch area in smaller RZs (by area, $\alpha = -0.897$, $P < 0.0001$; by perimeter, $\alpha = -0.868$, $P < 0.0001$) than in larger RZs. The ranks of RZs with regard to wetland habitat patches less than or equal to 2 ha within them indicated that all RZs have patches that were less than or equal to 2 ha, with the exception of RZs 31, 28, 22, and 2 (Table 6.14; Figure 6.26 and Figure 6.27). For wetland habitat patches less than or equal to 2 ha the mean original habitat patch area was weakly positively correlated with the area of RZ ($\alpha = 0.285$, $P = 0.0197$) and the perimeter of the RZ ($\alpha = 0.310$, $P = 0.0112$).

For wetland habitat patches less than or equal to 2 ha, the strongest positive correlation exists between total habitat patch count and the area of RZ ($\alpha = 0.909$, $P < 0.0001$) and the perimeter of the RZ ($\alpha = 0.905$, $P < 0.0001$). Because wetlands that were 2 ha or smaller were too small to reliably monitor with currently available satellite remote-sensing-derived land cover products, routine broad-scale monitoring of such small wetland areas may be difficult.

RZ gradient analyses in the LWRR suggest that larger habitat patches were relatively more likely to rebound (i.e., are less likely to be fragmented or

TABLE 6.12

Rank of 72 Federal Refuge Zones in Lower White River Region by Habitat Path Characteristics

		Rank of Federal Wildlife Refuge Zone Metric or Index (Zones 1–72)					
Patch Area (ha)	Patch Perimeter (km)	PIER	PSI	PCI	UPI	Relative Metric or Index Value Guide	Relative Vulnerability to Disturbance Guide
24	24	50	54	54	24	Lowest	Highest
50	40	24	40	52	40		
40	59	44	52	40	54		
59	60	59	24	24	60		
53	53	62	60	53	59		
60	50	47	53	60	53		
18	54	72	2	2	37		
51	37	8	59	59	50		
54	51	71	18	9	51		
8	18	20	9	18	9		
47	9	46	5	3	18		
44	8	53	3	42	2		
9	65	6	42	5	65		
6	6	67	1	51	3		
37	30	48	4	30	8		
30	47	40	30	33	47		
65	2	15	10	4	52		
4	4	26	6	10	6		
17	5	60	51	6	5		
15	3	4	33	1	30		
19	15	30	43	65	42		
26	44	69	15	37	34		
34	48	7	14	15	33		
3	7	17	65	8	48		
7	19	58	8	57	4		
71	34	45	7	7	7		
5	17	19	37	43	15		
2	33	43	36	14	19		
16	42	65	19	34	17		
1	1	18	23	19	44		
48	10	34	72	31	57		
10	72	1	57	48	10		
42	43	39	17	17	26		
67	58	36	34	23	1		

(continued)

TABLE 6.12 (CONTINUED)

Rank of 72 Federal Refuge Zones in Lower White River Region by Habitat Path Characteristics

		Rank of Federal Wildlife Refuge Zone Metric or Index (Zones 1–72)					
Patch Area (ha)	Patch Perimeter (km)	PIER	PSI	PCI	UPI	Relative Metric or Index Value Guide	Relative Vulnerability to Disturbance Guide
72	26	35	58	47	16		
14	71	10	48	72	43		
23	52	37	56	56	14		
43	14	9	62	36	58		
57	67	51	39	44	31		
58	23	5	31	50	23		
36	57	56	44	58	38		
33	16	21	67	39	71		
62	62	23	47	38	55		
39	36	61	20	62	39		
68	39	68	50	67	36		
55	55	70	38	16	67		
38	56	16	29	20	21		
46	38	66	69	55	72		
21	31	14	55	26	56		
45	21	3	16	29	62		
69	69	64	71	21	22		
22	68	22	26	71	69		
61	46	33	46	69	29		
31	22	11	21	11	11		
52	29	25	68	22	68		
56	20	54	11	68	13		
11	11	63	61	46	46		
70	61	42	22	12	12		
66	45	38	45	13	20		
29	70	2	49	27	70		
35	13	28	27	61	28		
13	12	55	12	25	45		
63	27	27	70	70	27		
28	28	57	13	49	61		
32	66	49	25	45	25		
12	25	29	66	28	41		
64	63	13	63	41	66		
27	35	32	28	66	63		

(continued)

TABLE 6.12 (CONTINUED)

Rank of 72 Federal Refuge Zones in Lower White River Region by Habitat Path Characteristics

						Rank of Federal Wildlife Refuge Zone Metric or Index (Zones 1–72)	
Patch Area (ha)	Patch Perimeter (km)	PIER	PSI	PCI	UPI	Relative Metric or Index Value Guide	Relative Vulnerability to Disturbance Guide
25	32	41	32	63	32		
20	41	31	41	32	35		
41	64	12	64	35	49		
49	49	52	35	64	64	Highest	Lowest

destroyed) than smaller habitat patches after disturbances, such as changes in hydrology, destruction of vegetation, or the establishment of opportunistic flora and fauna. However, large size alone may not be sufficient to ensure that a patch was capable of existing and flourishing in a disturbed setting; other contributing disturbance factors such as patch perimeter length, interior-to-edge ratio, sinuosity, and circularity may also be relevant. Additionally, human-induced disturbances, such as runoff, agriculture, and land conversion, may function as drivers of habitat degradation, fragmentation, and loss. Thus those areas with increased human-induced disturbance (as evidenced by present and future population density, population density change, and road development) in the LWRR may bring about future net wetland degradation or loss in the LWRR.

The positive correlation between wetland habitat area or the percent contribution of habitat area, with area of RZ indicating that larger refuges contain more wetland habitat than smaller refuges. However, the percent contribution of wetland habitat to a refuge was inversely correlated with the area of the refuge. The weakness of the correlation between RZ area and mean area of habitat patches within a RZ and outside a RZ (Table 6.10) indicated that the presence of a relatively large refuge was not necessarily a predictor of large habitat area within the refuge. The weak negative correlation between RZ area, and the *original area* of the habitat patches inside and outside the RZ; the habitat patch perimeter within a RZ; and the habitat patch interior-to-edge ratio within a RZ (Table 6.12) also indicated that relatively large available and suitable mallard duck winter habitat patches in the LWRR are not encompassed within current federal refuge boundaries.

The weak negative correlations between RZ area and human population density metrics in the LWRR indicated that larger RZs tend to exist in areas of lower human population densities, and smaller RZs tend to exist in areas of higher human population densities. Results of estimated future human

TABLE 6.13

Rank of 72 Federal Refuge Zones in Lower White River Region by Habitat Patch Human-Induced Disturbance Characteristics

								Rank of Federal Wildlife Refuge Zone Metric or Index (Zones 1–72)	
Rddens	Rdlen	PRI	Popdens 1990	Popdens 2011	Pdenchg 1990–2011	UHI	UVI	Relative Metric or Index Value Guide	Relative Vulnerability to Disturbance Guide
37	49	49	66	66	1	49	12	Highest	Highest
65	20	20	12	1	2	12	41		
56	12	12	63	2	66	13	57		
3	56	56	8	8	8	1	28		
19	13	13	64	12	20	64	52		
31	31	31	61	64	57	37	50		
69	37	37	13	61	64	31	49		
13	64	64	41	57	35	2	20		
72	58	58	43	41	45	41	56		
62	3	3	42	13	55	20	31		
49	70	70	57	63	32	3	63		
12	41	41	40	43	56	66	66		
10	11	11	31	42	61	56	1		
40	63	63	38	20	46	11	58		
8	1	1	49	45	41	8	64		
71	72	72	45	38	11	70	70		
34	69	69	44	62	10	19	37		
64	10	10	62	49	71	58	11		
58	19	19	2	47	14	25	8		
30	67	67	47	31	65	69	72		
39	62	62	70	40	17	21	69		
15	21	21	52	70	48	10	3		
48	66	66	1	52	7	71	10		
67	34	34	3	71	18	45	40		
1	71	71	71	44	28	57	67		
11	8	8	27	55	68	63	62		
41	27	27	58	58	16	72	21		
63	14	14	67	3	52	27	19		
18	45	45	46	67	47	14	27		
70	25	25	20	46	9	62	71		
46	61	61	5	35	51	39	14		
23	39	65	39	68	50	67	45		
17	65	39	30	32	53	65	61		
43	30	30	68	60	59	34	25		
21	23	23	60	28	60	61	65		

(continued)

TABLE 6.13 (CONTINUED)

Rank of 72 Federal Refuge Zones in Lower White River Region by Habitat Patch Human-Induced Disturbance Characteristics

				Rank of Federal Wildlife Refuge Zone Metric or Index (Zones 1–72)						
Rddens	Rdlen	PRI	Popdens 1990	Popdens 2011	Pdenchg 1990–2011	UHI	UVI	Relative Metric or Index Value Guide	Relative Vulnerability to Disturbance Guide
29	40	40	28	51	54	23	34		
20	15	15	55	39	24	29	2		
14	46	46	51	56	58	15	39		
4	17	17	36	65	69	46	30		
27	29	29	32	16	15	17	43		
66	43	43	56	5	70	18	46		
45	7	48	65	69	21	48	15		
6	48	7	34	27	62	35	23		
61	18	18	4	33	33	30	17		
7	6	6	26	72	25	55	29		
25	4	4	16	30	67	7	4		
32	32	35	69	34	39	32	13		
35	35	32	35	6	22	6	48		
28	28	28	33	36	19	28	7		
22	22	68	72	11	72	68	26		
68	68	22	6	17	6	16	18		
38	38	36	29	29	13	52	44		
55	55	55	37	25	29	47	6		
36	36	38	23	4	23	9	5		
16	16	16	25	37	49	51	36		
57	57	26	11	10	37	50	55		
52	52	57	17	7	42	53	38		
26	26	44	22	23	5	59	35		
42	42	33	10	15	12	60	42		
33	33	42	7	22	38	54	22		
44	44	5	15	48	34	24	32		
5	5	52	19	14	3	33	33		
2	2	47	21	21	36	40	68		
47	47	2	48	26	4	22	16		
9	9	9	14	19	43	43	47		
51	51	51	18	18	30	42	9		
54	54	50	9	9	44	5	51		
50	50	53	54	54	27	4	53		
53	53	59	50	50	26	38	59		
60	60	60	53	53	31	36	60		

(continued)

TABLE 6.13 (CONTINUED)

Rank of 72 Federal Refuge Zones in Lower White River Region by Habitat Patch Human-Induced Disturbance Characteristics

	Rank of Federal Wildlife Refuge Zone Metric or Index (Zones 1–72)

Rddens	Rdlen	PRI	Popdens 1990	Popdens 2011	Pdenchg 1990–2011	UHI	UVI	Relative Metric or Index Value Guide	Relative Vulnerability to Disturbance Guide
59	59	54	59	59	40	44	54		
24	24	24	24	24	63	26	24	Lowest	Lowest

population density change in the RZs (from 1990 to 2011) were inconclusive but could be expected to follow the same trend as the population density metrics, because large RZs in rural areas would likely lessen the effects of increased population density change.

The ranks of the four largest RZs (i.e., RZ 72, 67, 62, and 8) indicated that RZs were vulnerable to degradation as a result of smaller habitat patch size, less complex habitat patch shape, and human-induced disturbances within a habitat patch. All four of the largest RZs have a substantial proportion of very small wetland habitat patches within them (i.e., habitat ≤2 ha). Because of the high likelihood that these very small wetland habitat patches may be lost in the future (by definition), and because of the intrinsic difficulty in monitoring their loss, we recommended to decision makers that relatively small patches of wetland habitat in the LWRR be a high priority for future remote-sensing and field-based monitoring and conservation efforts.

Based on the number of human-induced disturbance and patch metric vulnerabilities among the four largest RZs, wetland habitat in RZ 8 was likely to experience the highest levels of disturbance and patch fragmentation or loss in the future, because its vulnerability to *all factors* except the unified vulnerability index is greater than the median. The next most vulnerable RZ among the four largest RZs are RZ 72 because it has greater than the median vulnerability to all factors except the unified patch index, population density factors, and the unified vulnerability index; and RZ 67 because it has greater than the median vulnerability to all factors except the patch perimeter metric, sinuosity index, circularity index, unified patch index, and population density factors. The least vulnerable among the four largest RZs is RZ 62 because it has greater than the median vulnerability to all factors except the patch area metric, patch perimeter metric, sinuosity index, circularity index, unified patch index, unified vulnerability index, and population density factors. Considering the relative total area of the four largest RZs (Table 6.15), the number of small habitat patches (i.e., ≤2 ha) in RZs 72, 67, and 62 was approximately proportional to each refuge's area, leaving RZ 8 with approximately half the expected number of small habitat patches. RZ

TABLE 6.14

Assessment of the Presence of Wetlands Less than or Equal to 2 ha Within Each of 68 Federal Refuge Zones in the Lower White River Region. Four Federal Refuge Zones (2, 22, 28, and 31) Did Not Contain Habitat Patches Less Than 2 ha.

Federal Refuge Zone ID	Mean Percent Contribution to Federal Zone Area (ha)	Mean Habitat Patch Area (ha)	Std. Dev. (ha)	Total Habitat Patch Count
72	0.000004	2814	3550	11391
8	0.000058	3110	3944	510
62	0.000081	2356	3157	460
67	0.000098	2563	3291	631
56	0.000120	2107	2741	552
29	0.000144	2343	2915	199
20	0.000168	1244	1816	184
15	0.000235	2961	3773	281
43	0.000258	2825	3306	182
23	0.000349	1400	1600	21
7	0.000380	2801	3502	237
58	0.000411	3157	4212	151
48	0.000423	1869	2661	76
4	0.000435	3293	4089	156
54	0.001000	508	795	18
37	0.001000	817	306	4
45	0.001000	931	1299	17
11	0.001000	1080	380	10
32	0.001000	1261	2154	33
60	0.001000	1415	2109	72
63	0.001000	1450	1755	36
68	0.001000	1462	1915	63
71	0.001000	1509	2152	154
46	0.001000	1747	1554	99
61	0.001000	1845	2513	75
18	0.001000	1880	2460	52
36	0.001000	2082	2777	33
69	0.001000	2139	2544	127
10	0.001000	2400	2578	64
30	0.001000	2600	2926	111
70	0.001000	2941	3859	36
6	0.001000	2950	3511	153
5	0.001000	3279	3627	86
49	0.001000	3460	4664	29
40	0.002000	1664	2289	27
66	0.002000	2163	2965	18

(continued)

TABLE 6.14 (CONTINUED)

Assessment of the Presence of Wetlands Less than or Equal to 2 ha Within Each of 68 Federal Refuge Zones in the Lower White River Region. Four Federal Refuge Zones (2, 22, 28, and 31) did not Contain Habitat Patches Less than 2 ha.

Federal Refuge Zone ID	Mean Percent Contribution to Federal Zone Area (ha)	Mean Habitat Patch Area (ha)	Std. Dev. (ha)	Total Habitat Patch Count
39	0.002000	2769	4365	37
27	0.002000	2931	4334	20
55	0.003000	640	247	5
19	0.003000	900	—	1
35	0.003000	1259	1848	28
16	0.003000	2006	1815	7
42	0.003000	2319	1765	4
44	0.003000	2696	3853	39
65	0.003000	3707	3967	48
14	0.004000	1319	—	1
50	0.004000	2482	3597	11
38	0.004000	2879	5049	23
34	0.004000	2893	3636	7
24	0.005000	805	862	3
59	0.005000	808	986	13
13	0.005000	1572	1958	8
17	0.006000	2399	2640	8
21	0.006000	4487	5975	7
1	0.006000	4519	5104	12
57	0.007000	1842	2133	12
33	0.007000	2645	3572	26
64	0.008000	649	823	4
47	0.009000	1684	1108	2
51	0.011000	2537	2922	3
12	0.012000	900	0	2
52	0.015000	4656	5776	4
3	0.015000	8593	—	1
41	0.016000	2590	2986	3
26	0.018000	2885	2594	3
25	0.019000	3022	2329	5
53	0.021000	1801	—	1
9	0.023000	3459	2695	5
Sum	0.282164	157082	174630	16701
Mean	0.004149	2310	2729	246
1 S.D.	0.005525	1228	1323	1378

TABLE 6.15

Assessment of presence of Wetlands Less Than or Equal to 2 ha within 68 Federal Refuge Zones in Lower White River Region

Federal Refuge Zone ID	Area (ha)	Perimeter (m)
12	7.6	1182
64	7.7	1237
2	8.0	1208
22	8.4	1211
53	8.4	1204
9	15.0	1548
59	15.1	2427
26	15.9	1596
25	16.0	1600
41	16.2	2025
31	16.3	1617
28	16.9	1646
24	17.7	1684
47	19.6	2112
51	22.3	1895
55	22.8	1899
57	28.1	2702
52	31.3	2353
14	33.7	2485
13	34.0	2475
19	35.3	2514
33	39.9	3799
17	40.0	2808
35	40.1	3823
54	40.3	2970
3	56.3	3132
16	65.0	4024
50	65.6	3773
37	66.5	3263
34	67.7	5196
21	78.5	4815
1	79.4	7156
11	80.8	3847
38	81.3	4032
42	89.4	5606
45	92.3	5224
40	95.1	4971
44	96.4	4822
63	99.5	5824

(continued)

TABLE 6.15 (CONTINUED)

Assessment of presence of Wetlands Less Than or Equal to 2 ha within 68 Federal Refuge Zones in Lower White River Region

Federal Refuge Zone ID	Area (ha)	Perimeter (m)
66	106.4	4380
32	114.2	4459
27	124.4	8113
65	128.7	4811
60	131.4	6252
68	161.1	7609
36	167.3	9596
39	174.3	10229
71	205.2	8099
10	212.4	7158
18	213.5	8143
30	230.7	10935
5	250.9	14060
46	251.4	9844
61	287.3	10260
69	295.1	11304
6	372.8	13889
23	401.8	10756
49	407.6	9278
48	441.6	9046
70	510.3	13622
7	738.1	25561
20	740.6	18248
4	757.3	25088
58	767.9	19509
43	1094.4	30733
15	1261.4	30217
29	1629.0	31502
56	1756.1	32308
67	2622.5	42974
62	2899.8	44889
8	5325.8	52555
72	64552.1	335315
Sum	89529.1	891387
Mean	2558.0	25468
1 S.D.	10838.3	55365

8 (Figure 6.26 and Figure 6.27) was a unit of the Bald Knob National Wildlife Refuge, a relatively recent wildlife refuge land acquisition, and was therefore predominated by agricultural land (Figure 6.26).

Thus RZ 8 was the largest federally owned parcel in the LWRR to have been recently (and almost completely) impacted by wetland fragmentation, loss, and human-induced disturbances. Therefore we recommended that the Bald Knob National Wildlife Refuge be the highest priority for wetland habitat restoration and protection among the four largest RZs in the LWRR.

Because indexes were (by definition) derived from other more directly measured metrics they may be less sensitive to the relative differences among patches than the metrics themselves (Yoder 1991; Karr and Chu 1997). We found that the indexes that included *patch characteristics* (i.e., the unified patch index and the unified vulnerability index) tended to rank the four largest RZs as less vulnerable than their component metrics. That is, the *patch characteristic* vulnerability *indices* used here tended to indicate that habitat was less vulnerable than the *component metrics* of that index. The purely *human-induced disturbance* indexes (i.e., the road index and the unified human index) tended to rank the four largest RZs consistently with their component metrics, with regard to the median parameter value.

Thus the results of the unified patch index and unified vulnerability index for RZ 72 (indicating that RZ 72 was relatively less vulnerable than other RZs) may be misleading because of this *dilution effect* of combining the sub-component metrics (Table 6.1). Accordingly, if index results were held aside, wetland vulnerability to landscape-ecological degradation factors in RZ 72 and 8 were similar. Consequently, we recommend to decision makers that RZ 72, the largest RZ, be given the second highest priority for wetland restoration and protection among the four largest RZs in the LWRR.

Results of the habitat vulnerability assessment of the RZ 67 index suggested that it was predominantly vulnerable to the influence of road construction and the presence of relatively smaller wetland habitat patches. Human-induced disturbance factors related to the presence of roads in RZ 67 may be partially mitigated by the robustness of patch characteristics within this RZ, with the exception of patch area. Therefore, we recommend to decision makers that RZ 67 be given the third highest priority for continued wetland restoration and protection among the four largest RZs in the LWRR.

Results of habitat vulnerability assessments for RZ 62 suggest that it was primarily vulnerable to the influence of roads. The human-induced disturbance factors related to the presence of roads in RZ 62 may be partially mitigated by the robustness of patch characteristics within this RZ, because all of the habitat size and shape metrics for RZ 62 were relatively high. Thus we recommend to decision makers that RZ 62 be given the lowest priority for continued wetland restoration and protection among the four largest RZs in the LWRR.

Utilizing Open Water and Riverine Conditions

Because the unified vulnerability index (UVI) integrates habitat patch size, habitat patch shape, and human-induced disturbance metrics into a single index value it is the most conservative (i.e., describes the *least potential impact* scenario) of the measures used to model habitat vulnerability. That is, a high UVI indicates component metrics that may range from low vulnerability to extreme vulnerability, but it encompasses all of the subcomponent metrics (see Table 6.12 and Table 6.13; discussion of using indexes versus component metrics in previous section). Thus the UVI was selected to model potential future land cover change in riparian wetland mallard duck winter habitat in a portion of RZ 72 (LWRR) given a hypothetical decrease in the extent and duration of riparian wetland flooding, given potential human development activities in the future. The hypothetical decrease in the extent and duration of flooding was assumed to involve a change from *permanently flooded* or *semipermanently flooded* wetland conditions to *intermittently flooded* or *rarely flooded* wetland conditions (after Cowardin et al. 1979; see Table 6.6 for mallard duck habitat hydrologic parameters). One hundred-thirteen kilometers of river channel and the surrounding landscape in the vicinity of the South Unit of the White River National Wildlife Refuge (Figure 6.28) was used to simulate these potential future hydrologic changes.

Results of a simulation along 113 km of (centerline) river and 226 km of (bank) riparian habitat (Figure 6.29a), per the mallard UVI determine the potential effects of a hypothesized future hydrologic change (Figure 6.29b), on wetlands within a 30 m region on the river's banks.

Mallard duck winter habitat under the simulated future hydrologic conditions indicated a decrease in the periodicity and duration of flooding along the 226 km of river-adjacent riparian wetlands, which could result in a conversion of 21% (2,822 ha) of the 13,514 ha of riparian wetland along the 113 km of White River system assessed (Figure 6.29b and Figure 6.29c). Accordingly, 79% of the assessed wetlands were relatively less vulnerable to conversion from such decreases in periodicity and the duration of flooding in the future.

The results of this gross simulation were substantially simplified because they described solely the conversion of wetlands from relatively *wetter conditions* to relatively *drier conditions*. Actual wetland change in riparian areas is likely more complex, involving many biophysical constraints (Vannote et al. 1980; Gregory et al. 1991; Middleton 1999). Additionally, because the UVI was a conservative model, hydrologic changes in this region of the LWRR may result in substantially greater biological effects, such that facultative-wetland or obligate-wetland plants (Reed 1988) might be less able to flourish, resulting in facultative or facultative-upland (Reed 1988) plant influx and establishment.

Thus future changes in the hydrology of the White River could result in the loss of plant species that are important resources for wetland organisms (e.g., *Potamogeton* spp. or *Polygonum* spp. for waterfowl forage). Improved hydrologic models for the White River could be used to improve upon the

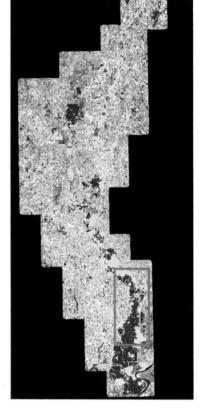

Mallard Winter habitat–unified
vulnerability index, patches > 2 hactares

■ < −3 Std. Dev. (Least vulnerability)
■ −3 − 2 Std. Dev.
▨ −2 − 1 Std. Dev.
▨ −1 − 0 Std. Dev.
  Mean
■ 0–1 Std. Dev.
  1–2 Std. Dev.
■ 2–3 Std. Dev.
■ > 3 Std. Dev. (Least vulnerability)

FIGURE 6.28
(See color insert.) The unified vulnerability index was applied to a hypothetical landscape
where change in the riparian wetland hydrology has occurred. Red rectangle indicates the
portion of the South Unit in the White River National Wildlife Refuge where landscape change
is hypothesized.

assumptions made in this simple model. Such improvements are part of the
detailed hydrologic assessments that are periodically investigated by the
USACE and may assist to determine the important linkages between hydrology and the vulnerability of biological resources for wetland organisms in
the LWRR.

The movement of water can cause a number of problems for management
of wetlands. In the past, many wetlands witnessed both unusual disasters
and unusual meteorological events that act as natural stressors and challenge
the integrity of wetlands (Figure 6.30). It is important to monitor wetlands
for usual wetland characteristics and conditions. Therefore, it is important to
have the capability to monitor following unusual events or even a disaster.

Remote measurements of water characteristics and conditions can be particularly helpful. Water appears much differently on imagery as compared to

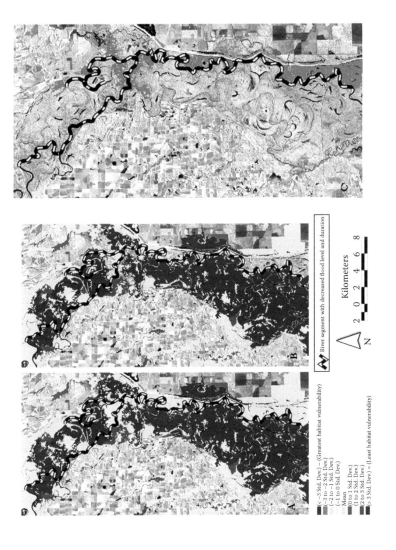

FIGURE 6.29

(See color insert.) (a) Mallard duck winter habitat vulnerability under current hydrologic conditions in the South Unit of the White River National Wildlife Refuge; (b) loss of mallard duck winter habitat under hypothetical decrease in flood stage and duration on the White River; (c) enlarged view of predicted net loss of mallard duck winter habitat under hypothetical decrease in flood stage and duration on the White River.

FIGURE 6.30
Flooding and other hydrologic anomalies are a key element in recognizing wetland functions in the landscape aided by recognition of the presence of dark-toned wetland plant residue and the light-toned surface of mineral soils.

bare soils and vegetation. During rainfall or flooding events, standing water and wet areas can be identified by the presence and abundance of water. These characteristics can assist in evaluating the results of flooding such as inundated areas and areas for potential risk.

Distribution and duration of flooding can be identified through use of operational remote sensing systems. The use of visible, near, and thermal infrared and RADAR sensors, and their measurements can supply the required data to address the continued chronic and catastrophic flooding that continues to plague the White River, Arkansas River, Ohio River, Missouri River, Mississippi River, and other associated tributaries. A current catastrophic flooding event impacted the populations of this area in 2011, and indications of flooding in the future are definitive.

We find that the abundance of landscape assessment approaches and the vast parameters involved have created a new area of exploration for wetland description. The approaches and techniques utilized in this chapter utilize the wealth of approaches and methods throughout this book, many of which are also currently being adapted to current (and future) environmental challenges, including the global impacts of climate change, broad impacts of flooding along rivers and throughout floodplains, and large-scale integrated ecosystem restoration.

Appendix: Mississippi River and Surrounding Landscape

Figures A.1 to A.7 are images of the Mississippi River and surrounding landscape, from its headwaters to the Gulf of Mexico. The Landsat image mosaic (Landsat ETM+ Panchromatic Band, 15 meter nominal spatial resolution) indicates selected features for reference of navigation and land use, and land cover in the floodplain.

Reference data is overlaid on the map series, as follows:

- River mile locations at 25 mile intervals (source: U.S. Army Corps of Engineers, St. Louis District)
- Lock and dam location (source: U.S. Army Corps of Engineers, St. Louis District)
- City location with state abbreviation
- Major Mississippi River tributaries
- Floodway locations (source: U.S. Army Corps of Engineers, St. Louis District)

The Mississippi River (Lake Itasca to Minneapolis, Minnesota)

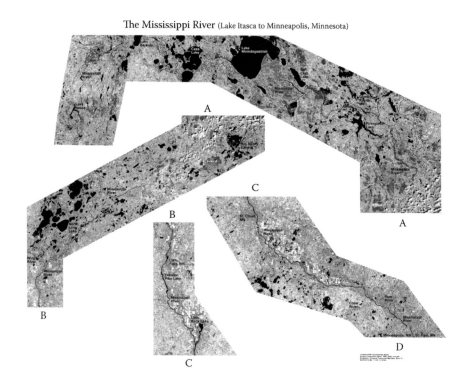

FIGURE A.1
Lake Itasca to Minneapolis, Minnesota.

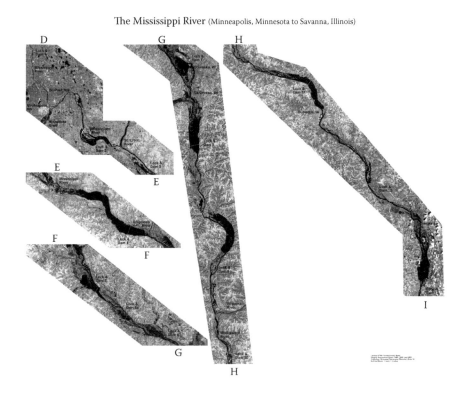

FIGURE A.2
Minneapolis, Minnesota, to Savanna, Illinois.

The Mississippi River (Clinton, Iowa to Lock & Dam 24)

FIGURE A.3
Clinton, Iowa, to Lock and Dam 24.

The Mississippi River (Lock & Dam 25 to Cairo, Illinois)

FIGURE A.4
Lock and Dam 25 to Cairo, Illinois.

The Mississippi River (Cairo, Illinois to Rosedale, Mississippi)

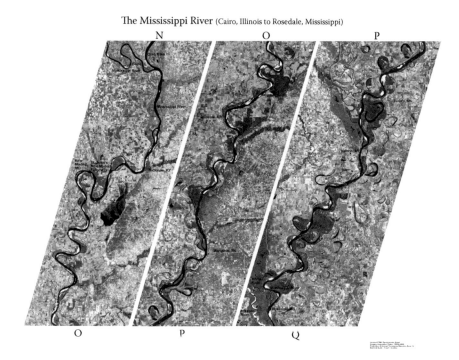

FIGURE A.5
Cairo, Illinois, to Rosedale, Mississippi.

The Mississippi River (Rosedale, Mississippi to St. Francisville, Louisiana)

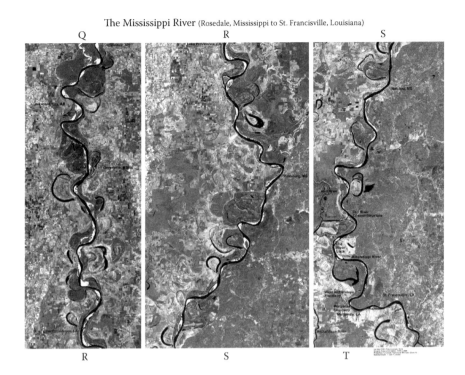

FIGURE A.6
Rosedale, Mississippi, to St. Francisville, Louisiana.

The Mississippi River (St. Francisville, Louisiana to the Gulf of Mexico)

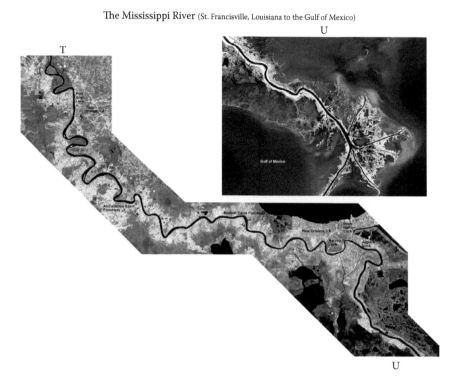

FIGURE A.7
St. Francisville, Louisiana, to the Gulf of Mexico.

References

Abood, S., A. Maclean, and L. Mason. 2012. Modeling riparian zones utilizing DEMS and flood height data. *Photogrammetric Engineering and Remote Sensing* 78:259–269.

Ackleson, S., and V. Klemas, 1987. Remote sensing of submerged aquatic vegetation in lower Chesapeake Bay: a comparison of Landsat MSS to TM imagery. Remote Sensing of Environment 22:235–248.

Adam, E., O. Mutanga, and D. Rugege. 2010. Multispectral and hyperspectral remote sensing for identification and mapping of wetland vegetation: A review. *Wetlands Ecology and Management* 18:281–296.

Adamus P., and K. Brandt. 1990. *Impacts on quality of inland wetlands of the United States: A survey of indicators, techniques, and applications of community level biomonitoring data*. EPA/600/3-90/073. Corvalis, OR: U.S. Environmental Protection Agency, Environmental Research Laboratory.

Adamus, P., L. Stockwell, E. Clairain, M. Morrow, L. Rozas, and R. Smith, 1991. *Wetland Evaluation Technique (WET)*. Technical Report WRP-DE-2. Vicksburg, MS: U.S. Army Waterways Experiment Station.

Akins, E., Y. Wang, and Y. Zhou. 2010. EO-1 Advanced Land Imager data in submerged aquatic vegetation mapping. In Y. Wang, *Remote sensing of coastal environments*. Boca Raton, FL: CRC Press.

Allen, A. 1987. *Habitat suitability index models: Mallard (winter habitat, Lower Mississippi Valley)*. Biological Report 82(10.132). Washington, DC: Department of Interior, U.S. Fish and Wildlife Service.

Alley, R., J. Marotzke, W. Nordhaus, J. Overpeck, D. Peteet, R. Pielke, R. Pierrehumbert, P. Rhines, T. Stocker, L. Talley, and J. Wallace. 2003. Abrupt climate change. *Science* 299:2005–2010.

American Society of Mechanical Engineers (ASME). 2008. *ASME water management technology vision and roadmap: Executive summary*. Washington, DC: American Society of Mechanical Engineers.

American Society of Photogrammetry and Remote Sensing. 1990. ASPRS accuracy standards for large-scale maps. *Photogrammetric Engineering and Remote Sensing* 56:1068–1070.

Amstrup, S.C., and J. Beecham. 1976. Activity patterns of radio-collared black bears in Idaho. *Journal of Wildlife Management* 40:340–348.

Anderson, J., E. Hardy, J. Roach, and R. Witmer. 1976. A land use classification system for use with remote-sensor data. U.S. Geological Survey Professional Paper 964. Washington, DC: U.S. Department of Interior.

Anderson, J., and J. Perry. 1996. Characterization of wetland plant stress using leaf reflectance spectra: Implications for wetlands remote sensing. *Wetlands* 16:477–487.

Andreas B.K. and Lichvar R.W. 1995. Floristic Index for Assessment Standards: A Case Study for Northern Ohio. Wetlands Research Program Technical Report WRP-DE-8. U.S. Army Corps of Engineers Waterways Experiment Station, Vicksburg, Mississippi, USA.

Anger, R. 2003. A soil survey enhancement of Landsat Thematic Mapper delineation of wetlands: A case study of Barry county. Master's thesis, Western Michigan University.

Antalovich, J. 2011. Disaster response aerial remote sensing following storms Irene and Lee. *Photogrammetric Engineering and Remote Sensing* 77:1185–1187.

Argialas, D., J. Lyon, and O. Mintzer. 1988. Quantitative description and classification of drainage patterns. *Photogrammetric Engineering and Remote Sensing* 54:505–509.

Arkansas Game and Fish Commission. 1998. *Black bear harvest report, 1997–1998.* Little Rock, AR: Arkansas Game and Fish Commission.

Artigas, F., and J. Yang. 2005. Hyperspectral remote sensing of marsh species and plant vigor gradient in the New Jersey Meadowlands. *International Journal of Remote Sensing* 26:5209–5220.

Backhaus, R., and B. Beule. 2005. Efficiency evaluation of satellite data products in environmental policy. *Space Policy* 21, 173–183.

Baker, C., R. Lawrence, C. Montagne, and D. Patten. 2006. Mapping wetlands and riparian areas using Landsat ETM+ imagery and decision-tree-based models. *Wetlands* 26: 465–474.

Baker, J.P., D.W. Hulse, S.V. Gregory, D. White, J. Van Sickle, P.A. Berger, D. Dole, and N.H. Schumaker. 2004. Alternative futures for the Willamette River Basin, Oregon. *Ecol. Appl.* 14, 313–324.

Baldassarre, G.A., R.J. Whyte, E.E. Quinlan, and E.G. Bolen. 1983. Dynamics and quality of waste corn available to postbreeding waterfowl in Texas. *Wildlife Society Bulletin* 11:25–31.

Ball, H., J. Jalava, T. King, L. Maynard, B. Potter, and T. Pulfer. 2003. *The Ontario Great Lakes wetland atlas.* Environment Canada and Ontario Ministry of Natural Resources, Canada.

Barbour, M., J. Burk, and W. Pitts. 1987. *Terrestrial plant ecology.* Menlo Park, CA: Benjamin/Cummings.

Barmore, W., and D. Stradley. 1971. Predation by black bear on mature elk. *Journal of Mammology* 52:199–202.

Barrette, J., P. August, and F. Golet. 2000. Accuracy assessment of wetland boundary delineation using aerial photography and digital orthophotography. *Photogrammetric Engineering and Remote Sensing* 66:409–416.

Bartonek, J., R. Blohm, R. Brace, F. Caswell, K. Gamble, H. Miller, R. Posahala, and M. Smith. 1984. Status and needs of the mallard. *Transactions of the North American Wildlife and Natural Resources Conference* 49:501–518.

Bastin, L., and C. Thomas. 1999. The distribution of plant species in urban vegetation fragments. *Landscape Ecology* 14:493–507.

Batema, D., G. Henderson, and L. Frederickson. 1985. Wetland invertebrate distribution in bottomland hardwoods as influenced by forest type and flooding regime. In *Proceedings of the Fifth Annual Hardwood Conference.* Urbana, IL: University of Illinois.

Bazzaz, F. 1986. Life history of colonizing plants. In H. A. Mooney and J. A. Drake, *Ecology of biological invasions of North America and Hawaii.* New York, NY: Springer-Verlag.

Becker, B., D. Lusch, and J. Qi. 2007. A classification-based assessment of the optimal spectral and spatial resolutions of coastal wetland imagery. *Remote Sensing of Environment* 108:111–120.

Beeman, L., and M. Pelton. 1980. Seasonal foods and feeding ecology of black bears in the Smoky Mountains. *Bear Biological Association Conference Series* 3:141–148.

Bellrose, F. 1976. *Ducks, geese, and swans of North America.* Harrisburg, PA: Stackpole Books.

Belluco, E., M. Camuffo, S. Ferrari, L. Modenese, S. Silvestri, A. Marani, and M. Marani. 2006. Mapping salt-marsh vegetation by multispectral and hyperspectral remote sensing. *Remote Sensing of Environment* 105:54–67.

Blom, C., G. Bogemann, P. Lann, A. Van der Sman, H. Van der Steeg, and L. Voesenek. 1990. Adaptations to flooding in plants from river areas. *Aquatic Botany* 38:29–47.

Boardman, J., and F. Kruse. 1994. Automated spectral analysis: A geological example using AVIRIS data, North Grapevine Mountains, Nevada. In *Proceedings of the Tenth Thematic Conference on Geologic Remote Sensing*. Ann Arbor, MI: Environmental Research Institute of Michigan.

Bolstad, P., and T. Lillesand. 1992. Rule-based classification models: Flexible integration of satellite imagery and thematic spatial data. *Photogrammetric Engineering and Remote Sensing* 58:965–971.

Bosch, W. 1978. A procedure for quantifying certain geomorphical features. *Geographical Analysis* 10:241–247.

Bray, H. 1974. Introductory remarks. *Proceedings of Eastern Black Bear Workshop* 2:7–10.

Brinson, M. 1993. *A hydrogeomorphic classification for wetlands*. Technical Report WRP-DE-4. Vicksburg, MS: U.S. Army Engineer Waterways Experimental Station.

Brinson, M., F. Hauer, L. Lee, W. Nutter, R. Rheinhardt, R. Smith, and D. Whigham. 1995. *A guidebook for application of hydrogeomorphic assessments to riverine wetlands*. Technical Report WRP-DE-11. Vicksburg, MS: U.S. Army Engineer Waterways Experiment Station.

Bromberg, S. 1990. Identifying ecological indicators: An environmental monitoring and assessment program. *Journal of the Air Pollution Control Association* 40:976–978.

Brown, D., E. Addink, J. Duh, and M. Bowersox. 2004. Assessing uncertainty in spatial landscape metrics derived from remote sensing data. In R. Lunetta and J. Lyon, *Remote sensing and GIS accuracy assessment*. Boca Raton, FL: CRC Press.

Brown, M., and J. Dinsmore. 1986. Implications of marsh size and isolation for marsh bird management. *Journal of Wildlife Management* 50:392–397.

Brown, S.L. 1984. The role of wetlands in the Green Swamp in: *Cypress Swamps* (K.C. Ewel, and H.T. Odum, eds.). University Presses of Florida, Gainesville, Florida.

Brown, S., Brinson, M. M., and Lugo, A. E. 1979. Structure and function of riparian wetlands. In: Johnson, R, R.; McCormick, J. F., tech. coord. Strategies for protection and management of floodplain wetlands and other riparian ecosystems, Proceedings of the symposium; 1978 December 11–13; Calloway Gardens, GA Gen, Tech. Rep. WO-12. Washington, D.C.: U.S. Department of Agriculture, Forest Service; 1979: 17–31.

Bryan, B.A., S. Hajkowicz, S. Marvanek, and M.D. Young. 2009. Mapping economic returns to agricultural for informing environmental policy in the Murray-Darling Basin, Australia. *Environ. Model. Assess.* 14, 375–390.

Bukata, R., J. Bruton, J. Jerome, and W. Haras. 1987. *A mathematical description of the effects of prolonged water level fluctuations on the areal extent of marshland*. Report RRB-87-02. Burlington, Ontario: Canada Centre for Inland Waters.

Bukata, R., J. Jerome, K. Kondratyev, and D. Pozdnyakov. 1995. *Optical properties and remote sensing of inland and coastal waters*. Boca Raton, FL: CRC Press.

Burnicki, A. 2011. Modeling the probability of misclassification in a map of land cover change. *Photogrammetric Engineering and Remote Sensing* 77:39–49.

Butera, K. 1983. Remote sensing of wetlands. *IEEE Transactions on Geoscience and Remote Sensing* GE-21:383–392.

Callan, O., and A. Mark. 2008. Using MODIS data to characterize seasonal inundation patterns in the Florida Everglades. *Remote Sensing of Environment* 112:4107–4119.

Cardoza, J. 1976. *The history and status of the black bear in Massachusetts and adjacent New England states*. Massachusetts Division of Fish and Wildlife Research Bulletin 18.

Carpenter, M. 1973. *The black bear in Virginia*. Virginia Commission of Game and Inland Fisheries.

Carreker, R. 1985. *Habitat Suitability Index models: Least tern*. Biological Report 82(10.103). Washington, DC: Department of Interior, U.S. Fish and Wildlife Service.

Carter, V. 1982. Applications of remote sensing to wetlands. In C. Johannsen and J. Sanders, *Remote sensing for resource management*. Ankeny, IA: Soil Conservation Society of America.

Carter, V. 1990. *Importance of hydrologic data for interpreting wetland maps and assessing wetland loss and mitigation*. Biology Report 90 (18). Washington, DC: Department of Interior, U.S. Fish and Wildlife Service.

Catling P., B. Freedman, C. Stewart, J. Kerekes, and L. Lefkovitch. 1986. Aquatic plants of acid lakes in Kejimkujik National Park, Nova Scotia; Floristic composition and relation to water chemistry. *Canadian Journal of Botany* 64:724–729.

Chander, G., B. Markham, and D. Helder. 2009. Summary of current radiometric calibration coefficients for Landsat MSS, TM, ETM+, and EO-1 ALI sensors. *Remote Sensing of Environment* 113:893–903.

Changnon, S. 2008. Assessment of flood losses in the United States. *Journal of Contemporary Water Research and Education* 138:38–44.

Chapin, F. 1991. Integrated response of plants to stress. *Bio-Science* 41: 29–36.

Chavez, P. 1988. An improved dark-object subtraction technique for atmospheric scattering correction of multispectral data. *Remote Sensing of the Environment* 24:459–479.

Christian, E. 2005. Planning for the Global Earth Observation System of Systems (GEOSS). *Space Policy* 21, 105–109.

Coffin, A.W. 2007. From roadkill to road ecology: A review of the ecological effects of roads. *J. Transp. Geogr.* 15, 396–406.

Congalton, R.G. 1988. A comparison of sampling schemes used in generating error matrices for assessing the accuracy of maps generated from remotely sensed data. Photogrammetric Engineering and Remote Sensing. 54:593–600.

Congalton, R. 1991. A review of assessing the accuracy of classifications of remotely sensed data. *Remote Sensing of Environment* 37:35–46.

Congalton, R. 1996. A quantitative comparison of change detection algorithms for monitoring eelgrass from remotely sensed data. Unpublished manuscript. Durham, NH: University of New Hampshire.

Congalton, R., and K. Green. 1998. *Accuracy assessment of remotely sensor data: Principles and practices*. Boca Raton, FL: CRC/Lewis Publishers.

Congalton, R., and K. Green. 2009. *Accuracy assessment of remotely sensor data: Principles and practices, 2nd Edition*. Boca Raton, FL: CRC/Lewis Publishers.

Connell, J., and R. Slatyer. 1977. Mechanisms of succession in natural communities and their role in community stability and organization. *American Naturalist* 111:1119–1144.

Constanza, R. 1980. Embodied energy and economic evaluation. *Science* 210:1219–1224.

Costa, M, O. Niemann, E. Novo, and F. Ahern. 2002. Biophysical properties and mapping of aquatic vegetation during the hydrological cycle of the Amazon floodplain using JERS-1 and Radarsat. *International Journal of Remote Sensing* 23:1401–1426.

Costlow, J., C. Boakout, and R. Monroe. 1960. The effect of salinity and temperature on larval development of *Sesarma cincereum* (Bosc.) reared in the laboratory. *Biological Bulletin* 118:183–202.

Cowan, I. 1972. The status and conservation of bears (Ursidae) of the world—1970. *International Conference of Bear Research and Management* 2:243–367.

Cowardin, L., V. Carter, F. Gollet, and E. LaRoe. 1979. *Classification of wetlands and deepwater habitats of the United States.* FWS/OBS–79/31. Washington, DC: Department of Interior, U.S. Fish and Wildlife Service.

Cox, W., and G. Cintron. 1997. The North American region in wetlands, biodiversity and the Ramsar convention. In A. Hails, *Proceedings of the Ramsar Convention on Wetlands.* http://www.ramsar.org/index_lib.htm.

Cushman, S., and K. McGarigal. 2004. Patterns in the species–environmental relationship depend on both scale and choice of response variables. *Oikos* 105:117–124.

Dahl, T. 1990. *Wetlands losses in the United States, 1780s to 1980s.* Washington, DC: Department of Interior, U.S. Fish and Wildlife Service.

Dahl, T. 2006. *Status and trends of wetlands in the conterminous United States 1998 to 2004.* Washington, DC: Department of Interior, U.S. Fish and Wildlife Service.

Dahl, T., and C. Johnson. 1991. *Status and trends of wetlands in the conterminous United States, mid-1970s to mid-1980s.* Washington, DC: Department of Interior, U.S. Fish and Wildlife Service.

Dahl, T., and M. Watmough. 2007. Current approaches to wetland status and trends monitoring in prairie Canada and the continental United States of America. *Canadian Journal of Remote Sensing* 33:S17–S27.

Davis, J. 1986. *Statistics and data analysis in geology.* 2nd ed. New York, NY: J. Wiley and Sons.

Day, J., T. Butler, and W. Conner. 1977. Productivity and nutrient export studies in a cypress swamp and lake system in Louisiana. In M. Wiley, ed., *Estuarine processes*, Vol. 2. New York, NY: Academic Press.

DeLaune, R., R. Boar, C. Lindau, and B. Kleiss. 1996. Denitrification in bottomland hardwood wetland soils of the Cache River. *Wetlands* 16:309–320.

de Leeuw, J., Y. Georgiadou, N. Kerle, A. de Gier, Y. Inoue, J. Ferwerda, M. Smies, and D. Narantuya. 2010. The function of remote sensing in support of environmental policy. *Remote Sens.* 2, 1731–1750.

De Roeck, E., N. Verhoest, M. Miya, H. Lievens, O. Batelaan, A. Thomas, and L. Brendonck. 2008. Remote sensing and wetland ecology: A South African case study. *Sensors* 8:3542–3556.

Diamond, J. 1974. Colonization of exploded volcanic islands by birds: The super tramp strategy. *Science* 184:803–806.

Dibble, E., J. Hoover, and M. Landin. 1995. *Comparison of abundance and diversity of young fishes and macroinvertebrates between two Lake Erie wetlands.* Technical Report WRP-RE-7. Vicksburg, MS: U.S. Army Engineer Waterways Experiment Station.

Dobson, J., and E. Bright. 1993. Large-area change analysis: The Coastwatch Change Analysis Project (CCAP). Sioux Falls, SD: 12th Pecora Symposium.

Dortch, M. 1996. Removal of solids, nitrogen, and phosphorus in the Cache River wetland. *Wetlands* 16:358–365.

Dwivedi, R., B. Rao, and T. Ravisankar. 2011. *Remote sensing of soils.* Boca Raton, FL: CRC Press.

Dunne, K., A. Rodrigo, and E. Samanns. 1998. *Engineering specification guidelines for wetland plant establishment and subgrade preparation.* Technical Report WRP-RE-19. Vicksburg, MS: U.S. Army Engineer Waterways Experiment Station.

Dzwonko, Z., and S. Loster. 1988. Species richness of small woodlands on the western Carpathian foothills. *Vegetatio* 76:15–27.

Economic and Social Research Institute (ESRI). 2007. *GIS for the conservation of woodlands and wetlands.* GIS Best Practices Series. Redlands, CA: ESRI.

Ehrenfeld, J. 1983. The effects of changes in land-use on swamps of the New Jersey Pine Barrens. *Biological Conservation* 25:253–275.

Ehrenfeld, J., and J. Schneider. 1991. *Chamaecyparis thyoides* wetlands and suburbanization: Effects on hydrology, water quality and plant community composition. *Journal of Applied Ecology* 28:467–490.

Elowe, K. 1984. Home range, movements, and habitat references of black bear (*Ursus americanus*) in Western Massachusetts. Master's thesis, University of Massachusetts.

Elton, C. 1958. *The ecology of invasions by animals and plants.* London, UK: Methuen.

Elvidge, C., D. Pack, E. Prins, E. Kihn, J. Kendall, and K. Baugh. 1998. Wildfire detection with meteorological satellite data: Results from New Mexico during June of 1996 using GOES, AVHRR, and DMSP-OLS. In R. Lunetta and C. Elvidge, *Remote sensing change detection.* Chelsea, MI, and Boca Raton, FL: Ann Arbor Press/CRC Lewis Publishers.

Engel-Cox, J.A., and R.M. Hoff. 2005. Science-policy data compact: Use of environmental monitoring data for air quality policy. *Environ. Sci. Policy* 8, 115–131.

Environment Canada. 1998. *Great Lakes fact sheet: How much habitat is enough?* Minister of Public Works and Government Services, Canada.

Environment Canada. 2002. *Great Lakes fact sheet: Great Lakes coastal wetlands—Science and conservation.* Canadian Wildlife Service (Ontario Region).

Evans, D., and H. Allen. 1995. *Mitigated wetlands restoration: Environmental effects at Green Bottom Wildlife management area, West Virginia.* Technical Report WRP-RE-10. Vicksburg, MS: U.S. Army Engineer Waterways Experiment Station.

Ewel, K. 1990. Swamps. In R. Myers, and J. Ewel, *Ecosystems of Florida.* Orlando, FL: University of Central Florida Press.

Falkner, E. 1994. *Aerial mapping: Methods and applications.* Boca Raton, FL: Lewis/CRC Press.

Falkner, E., and J. Morgan. 2001. *Aerial mapping: Methods and applications.* Boca Raton, FL: Lewis/CRC Press.

Fassnacht, K., W. Cohen, and T. Spies. 2006. Key issues in making and using satellite-based maps in ecology: A primer. *Forest Ecology and Management* 222:167–181.

Fauth, P., E. Gustafson, and K. Rabenold. 2000. Using landscape metrics to model source habitat for Neotropical migrants in the Midwestern U.S. *Landscape Ecology* 15:621–631.

Federal Emergency Management Agency. 1996. *Processed FEMAQ3 flood data 1:24,000, circa 1996*: Digital data. Washington, DC: Federal Emergency Management Agency.

Federal Geographic Data Committee. 1992. *Application of satellite data for mapping and monitoring wetlands.* Washington, DC: Wetlands Subcommittee.

Federal Geographic Data Committee. 2008. National Vegetation Classification Standard: Federal Geographic Data Committee website, http://www.fgdc.gov/standards/projects/FGDC-standards-projects/vegetation/index_html/. Accessed August 2012.

Federal Geographic Data Committee. 2010. Geospatial metadata: Federal Geographic Data Committee website, http://www.fgdc.gov/metadata/geospatial-meta-data-standards. Accessed August 2012.

Federal Interagency Committee for Wetlands Delineation. 1989. *Federal manual for identifying and delineating jurisdictional wetlands.* Washington, DC: U.S. Government Printing Office.

Fennessy M., R. Geho, B. Elifritz, and R. Lopez. 1998. *Testing the floristic quality assessment index as an indicator of riparian wetland quality.* Final Report to U.S. EPA. Columbus, OH: Ohio Environmental Protection Agency.

Fennessy M., M. Gray, R. Lopez, and M. Mack. 1998. *An assessment of wetlands using reference sites.* Final Report to U.S. EPA. Columbus, OH: Ohio Environmental Protection Agency.

Field, D., A. Reyer, C. Alexander, B. Shearer, and P. Genovese. 1990. *NOAA national coastal wetlands inventory.* Biology Report 90 (18). Washington, DC: Department of Interior, U.S. Fish and Wildlife Service.

Foody, G. 2002. Status of landcover classification accuracy assessment. *Remote Sensing of the Environment* 80:185–201.

Foody, G., and P. Atkinson, eds. 2002. *Uncertainty in remote sensing and GIS.* Hoboken, NJ: J. Wiley & Sons.

Forman, R. 1995. *Land mosaics: The ecology of landscapes and regions.* Cambridge, UK: Cambridge University Press.

Forman, R., A. Galli, and C. Leck. 1976. Forest size and avian diversity in New Jersey woodlots with some land use implication. *Oecologia* 26:1–8.

Forman, R., and M. Godron. 1986. *Landscape ecology.* New York, NY: J. Wiley & Sons.

Forman, R.T.T., D. Sperling, J.A. Bissonette, A.P. Clevenger, C.D. Cutshall, V.H. Dale, L. Fahrig, R.L. France, C.R. Goldman, K. Heanue, J. Jones, F. Swanson, T. Turrentine, and T.C. Winter. 2003. *Road Ecology: Science and Solutions*; Island Press: Washington, DC, USA.

Foth, H. D. 1990. *Fundamentals of soil science.* New York, NY: J. Wiley & Sons.

Frame, G. 1974. Black bear predation on salmon at Olson Creek, Alaska. *Zeitschrift fur Tierzuchtung und Zuchtungsbiologie* 35:23–38.

Fredrickson, L. 1979. *Floral and faunal changes in lowland hardwood forests in Missouri resulting from channelization, drainage, and impoundment.* FWS/OBS-78/91. Washington, DC: Department of Interior, U.S. Fish and Wildlife Service.

Frederickson, L., and L. Taylor. 1982. *Management of seasonally flooded impoundments for wildlife.* Resource Publication 148. Washington, DC: Department of Interior, U.S. Fish and Wildlife Service.

Frohn, R., R. Molly, C. Lane, and B. Autrey. 2009. Satellite remote sensing of isolated wetlands using object-oriented classification of Landsat-7 data. *Wetlands* 29:931–941.

Gacia, E., E. Ballesteros, L. Camarero, O. Delgado, A. Palau, J. Riera, and J. Catalan. 1994. Macrophytes from lakes in the eastern Pyrenees: Community composition and ordination in relation to environmental factors. *Freshwater Biology* 32:73–81.

Garbrecht, J., and L. Martz. 1993. Network and subwatershed parameters extracted from digital elevation models: The Bills Creek experience. *Water Resources Bulletin* 29:909–916.

Garofalo, D. 2003. Aerial photointerpretation of hazardous waste sites: An overview. In J. Lyon, *Geographic information system applications for watershed and water resources management.* London, UK: Taylor & Francis.

Garshelis, D., and M. Pelton. 1980. Activity of black bears in the Great Smoky Mountains National Park. *Journal of Mammology* 61:8–19.

Geographic Data Technology 1999. Data Integration. Lyme, New Hampshire.

Gesch, D., M. Oimoen, S. Greenlee, C. Nelson, M. Steuck, and D. Tyler. 2002. The National Elevation Dataset. *Photogrammetric Engineering & Remote Sensing.* 68:5–32.

Gilmore, M., D. Civco, E. Wilson, N. Barrett, S. Prisloe, J. Hurd, and C. Chadwick. 2010. Remote sensing and in situ measurements for delineation and assessment of coastal marshes and their constituent species. In J. Wang, *Remote sensing of coastal environment.* Boca Raton, FL: CRC Press.

Gilpin, M. 1981. Peninsula diversity patterns. *American Naturalist* 118:291–296.

Godwin, H. 1923. Dispersal of pond floras. *Journal of Ecology* 11:160–164.

Good, R., D. Whigham, and R. Simpson. 1978. *Freshwater wetlands: Ecological processes and management potential.* New York, NY: Academic Press.

Gorham, B. 1999. *Mapping Agricultural Landuse in the Mississippi Alluvial Valley of Arkansas: Report on the Mississippi Alluvial Valley of Arkansas Landuse/Landcover (MAVALULC) Project.* Center for Advanced Spatial Technologies, University of Arkansas, Fayetteville, Arkansas.

Gosselink, J., and R. Turner. 1978. The role of hydrology in freshwater wetland eco-systems. In R. Good, D. Whigham, and R. Simpson, eds., *Freshwater wetlands: Ecological processes and management potential.* New York, NY: Academic Press.

Government of Canada and GLNPO. 1995. *The Great Lakes: An environmental and resource book.* 3rd ed. Canada and U.S. Environmental Protection Agency. Toronto: Government of Canada.

Great Lakes Information Network (GLIN). 2004. The Great Lakes. http://www.great-lakes.net/lakes/. Accessed August 2012.

Great Lakes National Program Office (GLNPO). 1999. Selection of indicators for Great Lakes basin ecosystem health. In P. Bertram, v. 3. *State of the Lakes Ecosystem Conference.*

Green, A., M. Berman, P. Switzer, and M. Craig. 1988. A transformation for order-ing multispectral data in terms of image quality with implications for noise removal. *IEEE Transactions on Geoscience and Remote Sensing* 26:65–74.

Green, R. 1979. *Sampling design and statistical methods for environmental biologists.* New York, NY: J. Wiley & Sons.

Gregory, S., F. Swanson, W. McKee, and K. Cummins. 1991. An ecosystem perspective of riparian zones. *Bioscience* 41:540–551.

Grenier, M., Demers, A.-M., Labrecque, S., Benoit, M., Fournier, R.A., and Drolet, B. (2007). An object-based method to map wetland using RADARSAT-1 and Landsat-ETM images: test case on two sites in Quebec, Canada. Canadian jour-nal of remote sensing, 33 (Suppl.1), p. S28–S45.

Gross, M., M. Hardsky, V. Klemas, and P. Wolf. 1987. Quantification of biomass of the marsh grass *Spartina alterniflora loisel* using Landsat Thematic Mapper imagery. *Photogrammetric Engineering and Remote Sensing* 53:1577–1583.

Group on Earth Observations Geo Portal: GEOSS online; Available online: http://www.earthobservations.org/index.html (accessed on 10 March 2011).

Gucinski, H., M.J. Furniss, R.R. Ziemer, and M.H. Brookes. 2001. Forest Roads: A Synthesis of Scientific Information; General Technical Report PNW-GTR-509; Pacific Northwest Research Station, USDA Forest Service: Portland, OR, USA p. 103.

Guidice, J., and J. Ratti. 1995. *Interior wetlands of the United States: A review of wetland status, general ecology, biodiversity, and management.* Technical Report WRP-SM-9. Vicksburg, MS: U.S. Army Engineer Waterways Experiment Station.

Gustafson, E. 1998. Quantifying landscape spatial pattern: What is the state of the art? *Ecosystems* 1:143–156.

Gutzwiller, K., and S. Anderson. 1992. Interception of moving organisms; influences of path shape, size, and orientation on community structure. *Landscape Ecology* 6:293–303.

Hamazaki, T. 1996. Effects of patch shape on the number of organisms. *Landscape Ecology* 11:299–306.

Hamilton, R. 1978. Ecology of the black bear in southeastern North Carolina. Master's thesis, University of Georgia.

Hamilton, R., and R. Marchington. 1980. Denning and related activities of black bears in the coastal plain of North Carolina. *Bear Biology Association Conference Series* 3:121–126.

Hammer, D. 1992. *Creating freshwater wetlands.* Chelsea, MI: Lewis Publishers.

Harger, E. 1967. *Homing behavior of black bears.* Research Report No. 118. Lansing, MI: Michigan Department of Conservation.

Harken, J., and R. Sugumaran. 2005. Classification of Iowa wetlands using an airborne hyperspectral image: A comparison of the spectral angle mapper classifier and an object-oriented approach. *Canadian Journal of Remote Sensing* 31:167–174.

Harris, L. 1984. *The fragmented forest: Island biogeography theory and the preservation of biotic diversity.* Chicago, IL: University of Chicago Press.

Hay, A. 1979. Sampling design to test land-use map accuracy. *Photogrammetric Engineering and Remote Sensing* 45:529–533.

Hayes, D., J. Olin, J. Fischenich, and M. Palermo. 2000. *Wetlands engineering handbook.* ERDC/EL TR-WRP-RE-21. Vicksburg, MS: U.S. Army Engineer Research and Development Center.

Heber, M. 2008. *FGDC draft wetland mapping standard.* Washington, DC: Federal Geographic Data Committee, Wetland Subcommittee and Wetland Mapping Standards Workgroup.

Heggem, D., C. Edmonds, A. Neale, L. Bice, and K. Jones. 2000. An ecological assessment of the Louisiana Tensas River basin. *Environmental Monitoring and Assessment* 64:41–54.

Heilman, G.E., Jr., J.R. Strittholt, N.C. Slosser, and D.A. Dellasala. 2002. Forest fragmentation of the conterminous USA: Assessing forest intactness through road density and spatial characteristics. *BioScience* 52, 411–422.

Heitmeyer, M. 1985. Wintering strategies of female mallards related to dynamics of lowland hardwood wetlands in the upper Mississippi Delta. Doctoral dissertation, University of Missouri.

Heitmeyer, M., and L. Fredrickson. 1981. Do wetland conditions in the Mississippi delta hardwoods influence mallard recruitment? *Transactions of the North American Wildlife and Natural Resources Conference* 46:44–47.

Henderson, F., and A. Lewis. 2008. Radar detection of wetland ecosystems: A review. *International Journal of Remote Sensing* 29:5809–5835.

Herdendorf, C. 1987. *The ecology of the coastal marshes of western Lake Erie: A community profile.* Biological Report 85 (7.9). Washington, DC: Department of Interior, U.S. Fish and Wildlife Service.

Herdendorf, C., C. Raphael, E. Jaworski, and W. Duffy. 1986. *The ecology of Lake St. Clair wetlands: A community profile*, Biological Report 85 (7.7). Washington, DC: Department of Interior, U.S. Fish and Wildlife Service.

Hermy, M., and H. Stieperaere. 1981. An indirect gradient analysis of the ecological relationships between ancient and recent riverine woodlands to the south of Bruges (Flanders, Belgium). *Vegetatio* 44:43–49.

Herold, M., C.E. Woodcock, T.R. Loveland, J. Townshend, M. Brady, C. Steenmans, and C.C. Schmullius. 2008. Land-cover observations as part of a Global Earth Observation System of Systems (GEOSS): Progress, activities, and prospects. *IEEE Syst. J.* 2, 414–423.

Herrero, S. 1979. Black bears: The grizzly's replacement? In D. Burk, ed., *The black bear in modern North America*. Clinton, NJ: Boone and Crockett Club and Amwell Press.

Hess, L., J. Melack, S. Filoso, and Y. Wang. 1995. Delineation of inundated area and vegetation along the Amazon floodplain with the SIR-C Synthetic Aperture Radar. *IEEE Transactions on Geoscience and Remote Sensing* 33:896–903.

Hess, L., J. Melack, and D. Simonett. 1990. Radar detection of flooding beneath the forest canopy: A review. *International Journal of Remote Sensing* 11:1313–1325.

Hey, D., and N. Phillippi, 1995. Flood reduction through wetland restoration: The upper Mississippi as a case study. *Restoration Ecology* 3: 4–17.

Hirano, A., M. Madden, and R. Welch. 2003. Hyperspectral image data for mapping wetland vegetation. *Wetlands* 23:436–448.

Hook, D., B. Davis, J. Scott, J. Struble, C. Bunton, and E. Nelson. 1995. Locating delineated wetland boundaries in coastal South Carolina using global positioning systems. *Wetlands* 15:31–36.

Hopkinson, C.S., Jr., and J.W. Day, Jr. 1980a. Modeling the relationship between development and storm water and nutrient runoff. *Environmental Management.* 4:315–324.

Howard, J. 1970. *Aerial photo-ecology.* New York, NY: American Elsevier.

Hruby, T., W. Cesanek, and K. Miller. 1995. Estimating relative wetland values for regional planning. *Wetlands* 15:93–107.

Huadong, G. 2010. *Atlas of the remote sensing of the Wenchuan earthquake.* Boca Raton, FL: CRC Press.

Huang, C., L. Yang, C. Homer, B. Wylie, J. Vogelman, and T. DeFelice. 2001. *At-sensor reflectance: A first order normalization of Landsat 7 ETM+ images.* Sioux Falls, SD: U.S. Geological Survey White Paper.

Huang, P., Y. Li, and M. Sumner. 2011. *Handbook of soil sciences.* 2nd ed. Boca Raton, FL: CRC Press.

Hugie, R. 1979. Working group report: Central and northeast Canada and United States. In D. Burk, ed., *The black bear in modern North America*. Clinton, NJ: Boone and Crockett Club and Amwell Press.

Hunsaker, C., and D. Carpenter. 1990. *Environmental monitoring and assessment program—Ecological Indicators.* EPA 600/3-90/060. Research Triangle Park, N.C.: U. S. Environmental Protection Agency.

Hupp, C., and D. Bazemore. 1993. Temporal and spatial patterns of wetland sedimentation, West Tennessee. *Journal of Hydrology* 141:179–196.

Hupp, C., and E. Morris. 1990. A dendrogeomorphic approach to measurement of sedimentation in a forested wetland, Black Swamp, Arkansas. *Wetlands* 10:107–124.

Hutchinson, G. 1975. *Limnological botany. Volume 3 of a treatise on limnology.* New York, NY: J. Wiley & Sons.

Islam, M, P. Thenkabail, R. Kulawardana, R. Alankara, S. Gunasinghe, C. Edussriya, and A. Gunawardana. 2008. Semi-automated methods for mapping wetlands using Landsat ETM+ and SRTM data. *International Journal of Remote Sensing* 29:7077–7106.

Jakubauskas, M., K. Kindscher, A. Fraser, D. Debinski, and K. Price. 2000. Close-range remote sensing of aquatic macrophyte vegetation cover. *International Journal of Remote Sensing* 21:3533–3538.

Jehl, D. 2002. Arkansas rice farmers run dry, and U.S. remedy sets off debate. *New York Times*, November 11.

Jenning, M. 1995. Gap analysis today: A confluence of biology, ecology, and geography for management of biological resources. *Wildlife Society Bulletin* 23:658–662.

Jensen, J. 1996. *Introductory digital image processing*. Upper Saddle River, NJ: Prentice Hall.

Jensen, J. 2004. *Introductory digital image processing: A remote sensing perspective*. Englewood Cliffs, NJ: Prentice Hall.

Jensen, J. 2006. *Remote sensing of the environment: An earth resource perspective*. Englewood Cliffs, NJ: Prentice Hall.

Jensen, J., S. Narumaiani, O. Weatherbee, and H. Mackey. 1993. Measurement of seasonal and yearly cattail and waterlily changes using multidate SPOT panchromatic data. *Photogrammetric Engineering and Remote Sensing* 59:519–525.

Jenson, S., and J. Domingue. 1988. Extracting topographic structure from digital elevation data for geographic information system analysis. *Photogrammetric Engineering and Remote Sensing* 54:1593–1600.

Ji, W. 2007. *Wetland and water resource modeling and assessment: A watershed perspective*. Boca Raton, FL: CRC Press.

Ji, W., and L. Mitchell. 1995. Analytical model-based decision support GIS for wetland resource management. In J. Lyon and J. McCarthy, *Wetland and environmental applications of GIS*. Boca Raton, FL: Lewis Publishers.

Johansen, K., L. Arroyo, S. Phinn, and C. Witte. 2010. Comparison of geo-object base and pixel-based change detection of riparian environments using high spatial resolution multi-spectral imagery. *Photogrammetric Engineering and Remote Sensing* 76:123–136.

Johnston, C. 1989. Human impacts to Minnesota wetlands. *Journal of the Minnesota Academy of Science* 55:120–124.

Johnston, C. 1994. Cumulative impacts to wetlands. *Wetlands* 14:49–55.

Johnston, C., N. Detenbeck, and G. Niemi. 1990. The cumulative effect of wetlands on stream water quality and quantity, a landscape approach. *Biogeochemistry* 10:105–141.

Johnston, J., and L. Handley. 1990. *Coastal mapping programs at the U.S. Fish and Wildlife Service's National Wetlands Research Center*. Biology Report 90 (18). Washington, DC: Department of Interior, U.S. Fish and Wildlife Service.

Jones, J. R., M. F. Knowlton, D. V. Obrecht, and E. A. Cook. 2004. Importance of landscape variables and morphology on nutrients in Missouri reservoirs. *Canadian Journal of Fisheries and Aquatic Sciences* 61:1503-1512.

Jones, K., A. Neale, M. Nash, R. Van Remortel, J. Wickham, K. Riitters, and R. Odum. 1985. Trends expected in stressed ecosystems. *Bioscience* 35:419–422.

Jones, K., A. Neale, M. Nash, K. Riitters, J. Wickham, R. O'Neill, and R. Van Remortel. 2000. Landscape correlates of breeding bird richness across the United States mid-Atlantic region. *Environmental Monitoring and Assessment* 63:159–174.

Jones, K., A. Neale, M. Nash, R. Van Remortel, J. Wickham, K. Riitters, and R. O'Neill. 2001. Predicting nutrient and sediment loadings to streams from landscape metrics: A multiple watershed study from the United State mid-Atlantic region. *Landscape Ecology* 16:301–312.

Jones, K., K. Riiters, J. Wickham, R. Tankersley, R. O'Neill, D. Chaloud, E. Smith, and A. Neale. 1997. *An ecological assessment of the United States mid-Atlantic region: A landscape atlas.* EPA/600/R-97/130. U.S. Environmental Protection Agency, Office of Research and Development.

Jonkel, C., and I. Cowan. 1971. The black bear in the spruce-fir forest. *Wildlife Monographs* 27:1–57.

Jorde, D., G. Krapu, and R. Crawford. 1983. Feeding ecology of mallards wintering in Nebraska. *Journal Wildlife Management* 47:1044–1053.

Jorde, D., G. Krapu, R. Crawford, and M. Hay. 1984. Effects of weather on habitat selection and behavior of mallards wintering in Nebraska. *Condor* 86:258–265.

Juracek, F., A. Perry, and E. Putnam. 2001. *The 1951 floods in Kansas revisited.* USGS Fact Sheet 041-01. U.S. Geological Survey.

Kadlec, R., and S. Wallace. 2009. *Treatment wetlands.* Boca Raton, FL: CRC Press.

Kadmon, R., and R. Harari-Kremer. 1999. Long-term vegetation dynamics using digital processing of historical aerial photographs. *Remote Sensing of Environment* 68:164–176.

Kampe, T.U., B.R. Johnson, M. Kuester, and M. Keller. 2010. NEON: The first continental-scale ecological observatory with airborne remote sensing of vegetation canopy biochemistry and structure. *J. Appl. Remote Sens.* 4, 043510.

Kang, Y., T. Zahniser, L. Wolfson, and J. Bartholic. 1994. *WIMS: A prototype wetlands information management system for facilitating wetland decision making.* Reno, NV: The Annual ACSM/ASPRS Convention.

Karr, J., and E. Chu. 1997. *Biological monitoring and assessment: Using multimetric indexed effectively.* EPA/235/R97/001. Seattle, WA: University of Washington.

Kasischke, E., J. Melack, and M. Dobson. 1997. The use of imaging radars for ecological applications: A review. *Remote Sensing of the Environment* 59:141–156.

Kasischke, E., K. Smith, L. Bourgeau-Chavez, E. Romanowicz, S. Brunzell, and C. Richardson. 2003. Effects of seasonal hydrologic patterns in south Florida wetlands on radar backscatter measured from ERS-2 SAR imagery. *Remote Sensing of Environment* 88:423–444.

Kauth, R., and G. Thomas. 1976. Tasseled cap—a graphic description of the spectral-temporal development of agricultural crops as seen by Landsat. *Proceedings from Remotely Sensed Data Symposium.* West Lafayette, IN: Purdue University.

Kellman, M. 1996. Redefining roles: Plant community reorganization and species preservation in fragmented systems. *Global Ecology and Biogeography Letters* 5:111–116.

Keough, J., T. Thompson, G. Guntenspergen, and D. Wilcox. 1999. Hydrogeomorphic factors and ecosystem responses in coastal wetlands of the Great Lakes. *Wetlands* 19:821–834.

Kilgore, K., and J. Baker. 1996. Patterns of larval fish abundance in a bottomland hardwood wetland. *Wetlands* 16:288–295.

Kitchens, W., J. Dean, L Stevenson, and J. Cooper. 1975. The Santee Swamp as a nutrient sink. In F. Howell, J. Gentry, and M. Smith, eds., *Mineral cycling in southeastern ecosystems.* Washington, DC: Energy Research and Development Administration Symposium Series.

Kleiss, B. 1996. Sediment retention in a bottomland hardwood wetland in eastern Arkansas. *Wetlands* 16:321–333.

Klemas, V. 2009. Sensors and techniques for observing coastal ecosystems. In X. Yang, *Remote sensing and geospatial technologies for coastal ecosystem assessment and management*. Berlin, Germany: Springer-Verlag.

Klemas, V. 2011. Remote sensing of wetlands: Case studies comparing practical techniques. *Journal of Coastal Research* 27:418–427.

Klepinger, K., and H. Norton. 1983. Black bear. In E. Deems and D. Pursley, eds., *North American furbearers: A contemporary reference*. International Association of Fish and Wildlife Agencies, and Maryland Department of Natural Resources.

Koren, I., L.A. Remer, and K. Longo. 2007. Reversal of trend of biomass burning in the Amazon. *Geophys. Res. Lett.* 34, L20404.

Krapu, G. 1981. The role of nutrient reserves in mallard reproduction. *Auk* 98:29–38.

Kress, M., M. Graves, and S. Bourne. 1996. Loss of bottomland hardwood forests and forested wetlands in the Cache River Basin, Arkansas. *Wetlands* 16:258–263.

Kushwaha, S., R. Dwivedi, and B. Rao. 2000. Evaluation of various digital image processing techniques for detection of coastal wetlands using ERS-1 SAR data. *International Journal of Remote Sensing* 21:565–579.

Lal, R., and B. Stewart. 2012. *World soil resources and food security.* Boca Raton, FL: CRC Press.

Lampman, J. 1993. *Bibliography of remote sensing techniques used in wetland research.* Springfield, VA: National Technical Information Service.

Landers, J., R. Hamilton, A. Johnson, and R. Marchington. 1979. Foods and habitat of black bears in southeastern North Carolina. *Journal of Wildlife Management* 43:143–153.

Lang, M., E. Kasischke, S. Prince, and K. Pittman. 2008. Assessment of C-band synthetic aperture radar data for mapping and monitoring coastal plain forested wetlands in the Mid-Atlantic region, U.S.A. *Remote Sensing of Environment* 112:4120–4130.

Lautenbacher, C.C. 2006. The Global Earth Observation System of Systems: Science serving society. *Space Policy* 22, 8–11.

LeCount, A. 1980. Some aspects of black bear ecology in the Arizona chaparral. *Bear biological association conference series* 3:175-179.

Lee, C., and S. Marsh. 1995. The use of archival Landsat MSS and ancillary data in a GIS environment to map historical change in an urban riparian habitat. *Photogrammetric Engineering and Remote Sensing* 61:999–1008.

Lee, K., and R. Lunetta. 1990. *Watershed characterization using Landsat Thematic Mapper (TM) satellite imagery, Lake Pend Oreille, Idaho.* Report TS-AMD-90C10. Las Vegas, Nev.: U.S. Environmental Protection Agency.

Lee, K., and R. Lunetta. 1995. Wetland detection methods investigation. In Lyon and McCarthy.

Leibowitz, S., B. Abbruzzese, P. Adamus, L. Hughes, and J. Irish. 1992. *A synoptic approach to cumulative impact assessment: A proposed methodology.* EPA/600/R-92/167. Corvalis, Ore.: U.S. Environmental Protection Agency.

Leibowitz, S. and T. Nadeau. 2003. Isolated wetlands: State-of-the-science and future directions. *Wetlands* 23:662-683.

Leonard, S., C. Bishop, and A. Gendron. 2000. Amphibians and reptiles in Great Lakes wetlands: Threat and conservation. http://www.on.ec.gc.ca/wildlife/factsheets/fs_amphibians-e.html

Lewis, W. 1995. *Wetlands characteristics and boundaries*. Washington, DC: National Academy Press.

Lichvar, R., and J. Wakeley. 2004. *Review of ordinary high water mark indicators for delineating arid streams in the Southwestern United States*. ERDC TR-04-1. Hanover, N.H.: U.S. Army Engineer Research and Development Center, Cold Regions Research and Engineering Laboratory.

Lillesand, T., and R. Kiefer. 1994. *Remote sensing and image interpretation*. New York, NY: J. Wiley & Sons.

Lillesand, T., R. Kiefer, and J. Chipman. 2004. *Remote sensing and image interpretation*. New York, NY: J. Wiley & Sons.

Lindberg, W., J. Persson, and S. Wold. 1983. Partial least-square method for spectrofluorimetric analysis of mixture of humic acid and lignisulfonate. *Analytical Chemistry* 55:643–648.

Linderman, M., Y. Zeng, and P. Rowhani. 2010. Climate and land-use effects on interannual fAPAR variability from MODIS 250 m data. *Photogrammetric Engineering and Remote Sensing* 76:807–816.

Lindzey, J., W. Kordek, G. Matula, and W. Piekielek. 1976. The black bear in Pennsylvania-status movements, values, and management. In M. Pelton, G. Folk, and J. Lentfer, eds., *Bears—Their biology and management; Proceedings of the 3rd International Conference on Bear Research and Management*.

Linsley, R., and J. Franzini. 1979. *Water resources engineering*. New York, NY: McGraw-Hill.

Llewellyn, D.W., G.P. Shaffer, N.J. Craig, L. Creasman, D. Pashley, M. Swan, and C. Brown.1996. A decision-support system for prioritizing restoration sites on the Mississippi River alluvial plain. *Conservation Biology* 10(5): 1446–1455.

Long, K., and J. Nestler. 1996. Hydroperiod changes as clues to impacts. *Wetlands* 16:258–263.

Lopez, R. 2006. *An ecological assessment of invasive and aggressive plant species in coastal wetlands of the Laurentian Great Lakes: A combined field-based and remote sensing approach*. Las Vegas, NV: U.S. Environmental Protection Agency, Environmental Sciences Division.

Lopez, R., C. Davis, and M. Fennessy. 2002. Ecological relationships between landscape change and plant guilds in depressional wetlands. *Landscape Ecology* 17:43–56.

Lopez, R., D. Heggem, C. Edmonds, K. Jones, L. Bice, M. Hamilton, E. Evanson, C. Cross, and D. Ebert. 2003. *A landscape case study of ecological vulnerability: Arkansas' White River watershed and the Mississippi alluvial valley ecoregion*. EPA/600/R-03/057. Washington, DC: United States Environmental Protection Agency.

Lopez, R., D. Heggem, D. Sutton, T. Ehil, R. Van Remortel, E. Evanson, and L. Bice. 2006. *Using landscape metrics to develop indicators of Great Lakes coastal wetland condition*. EPA/X-06/002. Las Vegas, NV: U.S. Environmental Protection Agency, Environmental Sciences Division.

Lopez, R., M. Nash, D. Heggem, L. Bice, E. Evanson, L. Woods, R. Van Remortel, M. Jackson, D. Ebert, and T. Harris. 2006. *Water quality vulnerability in the Ozarks using landscape ecology metrics: Upper White River browser (v2.0)*. EPA/600/C-06/017. Washington, DC: United States Environmental Protection Agency.

Lopez, R.D., Nash, M.S., Heggem, D.T., and D.W. Ebert. 2008. Watershed Vulnerability Predictions for the Ozarks using Landscape Metrics. *Journal of Environmental Quality*. 37:1769–1780.

Lory, J. 1999. *Agricultural phosphorus and water quality*, Publication G9181. Columbia, MO.: University of Missouri.

Loveland, T. and D. Ohlen, 1993. Experimental AVHRR land data sets for environmental monitoring and modeling. In M. Goodchild, B. Parks, and L. Steyaert, eds., *Environmental modeling with GIS*. New York, NY: Oxford University Press.

Lugo, A.E., and H. Gucinski. 2000. Function, effects, and management of forest roads. *Forest Ecol. Manag.* 133, 249–262.

Lunetta, R., M. Balogh, and J. Merchant. 1999. Application of multi-temporal Landsat 5 TM imagery for wetland identification. *Photogrammetric Engineering and Remote Sensing* 65:1303–1310.

Lunetta, R., R. Congalton, L. Fenstermaker, J. Jensen, K. McGwire, and L. Tinney. 1991. Remote sensing and geographic information system data integration: Error sources and research issues. *Photogrammetric Engineering and Remote Sensing* 57:677–688.

Lunetta, R., and C. Elvidge. 1998. *Remote sensing change detection.* Chelsea, MI, and Boca Raton, FL: Ann Arbor Press/CRC Lewis Publishers.

Lunetta, R., J. Knight, H. Paerl, J. Streicher, B. Pierls, T. Gallo, J. Lyon, T. Mace, and C. Buzzelli. 2009. Measurement of water colour using AVIRIS imagery to assess the potential for an operational monitoring capability in the Pamlico Sound estuary, U.S.A. *International Journal of Remote Sensing* 30: 3291–3314.

Lunetta, R., and J. Lyon. 2004. *Remote sensing and GIS accuracy assessment.* Boca Raton, FL: CRC Press.

Lunetta, R., J. Lyon, C. Elvidge, and B. Guindon. 1998. North American landscape characterization: Dataset development and data fusion issues. *Photogrammetric Engineering and Remote Sensing* 64:821–829.

Lunetta, R., J. Lyon, D. Worthy, J. Sturdevant, J. Dwyer, D. Yuan, C. Elvidge, and L. Fenstermaker. 1993. *North American Landscape Characterization (NALC), Landsat Pathfinder technical plan.* EPA 600/X-93/009. Las Vegas, NV: U.S. Environmental Protection Agency.

Lunetta, R., Y. Shao, J. Ediriwickrema, and J. Lyon. 2010. Monitoring agricultural cropping patterns across the Laurentian Great Lakes Basin using MODIS-NDVI data. *International Journal of Applied Earth Observations and Geoinformation* 12:81–88.

Luoto, M. 2000. Modeling of rare plant species richness by landscape variable in an agricultural area in Finland. *Plant Ecology* 149:157–168.

Lyon, J. 1979. Remote sensing of coastal wetlands and habitat quality of the St. Clair Flats, Michigan. 13th International Symposium on Remote Sensing of Environment, Ann Arbor, Michigan.

Lyon, J. 1980. Data sources for analyses of Great Lakes wetlands. Annual Meeting of the American Society for Photogrammetry, St. Louis, MO.

Lyon, J. 1981. The influence of Lake Michigan water levels on wetland soils and distribution of plants in the Straits of Mackinac, Michigan. Doctoral dissertation, University of Michigan.

Lyon, J. 1987. Maps, aerial photographs and remote sensor data for practical evaluations of hazardous waste sites. *Photogrammetric Engineering and Remote Sensing* 53:515–519.

Lyon, J. 1993. *Practical handbook for wetland identification and delineation.* Boca Raton, FL: CRC Press/Lewis Publishers.

Lyon, J. 1995. Wetlands: How to avoid getting soaked. *Professional Surveyor* 15:16–18.

Lyon, J. 2001. *Wetland landscape characterization: GIS, Remote sensing, and image analysis.* Chelsea, MI: Ann Arbor Press.

Lyon, J. 2003. *Geographic information system applications for watershed and water resources management.* London, U.K.: Taylor & Francis.

Lyon, J., and K. Adkins. 1995. Use of a GIS for wetland identification, the St. Clair Flats, Michigan. In J. Lyon and J. McCarthy, *Wetland and environmental applications of GIS.* Boca Raton, FL: Lewis Publishers.

Lyon, J., K. Bedford, J. Yen Chien-Ching, D. Lee, and D. Mark. 1988. Suspended sediment concentrations as measured from multidate Landsat and AVHRR data. *Remote Sensing of Environment* 25:107–115.

Lyon, J., and R. Drobney. 1984. Lake level effects as measured from aerial photos. *Journal of Surveying Engineering* 110:103–111.

Lyon, J., R. Drobney, and C. Olson. 1986. Effects of Lake Michigan water levels on wetland soil chemistry and distribution of plants in the Straits of Mackinac. *Journal of Great Lakes Research* 12:175–183.

Lyon, J., E. Falkner, and W. Bergen. 1995. Cost estimating photogrammetric and aerial photography services. *Journal of Surveying Engineering* 121:63–86.

Lyon, J., and R. Greene. 1992. Lake Erie water level effects on wetlands as measured from aerial photographs. *Photogrammetric Engineering and Remote Sensing* 58:1355–1360.

Lyon, J., J. Heinen, R. Mead, and N. Roller. 1987. Spatial data for modeling wildlife habitat. *Journal of Surveying Engineering* 113:88–100.

Lyon, J., and W. Hutchinson. 1995. Application of a radiometric model for evaluation of water depths and verification of results with airborne scanner data. *Photogrammetric Engineering and Remote Sensing* 61:161–166.

Lyon, J., R. Lunetta, and D. Williams. 1992. Airborne multispectral scanner data for evaluation of bottom types and water depths of the St. Marys River, Michigan. *Photogrammetric Engineering and Remote Sensing* 58:951–956.

Lyon, J., and L. Lyon. 2011. *Practical handbook for wetland identification and delineation.* 2nd ed. Boca Raton, FL: CRC Press.

Lyon, J., and J. McCarthy. 1981. SEASAT radar imagery for detection of coastal wetlands. 15th International Symposium on Remote Sensing of Environment, Ann Arbor, Michigan.

Lyon, J., and J. McCarthy. 1995. *Wetland and environmental applications of GIS.* Boca Raton, FL: Lewis Publishers.

Lyon, J., and C. Olson. 1983. *Inventory of coastal wetlands.* Ann Arbor, MI.: Michigan Sea Grant Program Publication, University of Michigan.

Lyon, J., D. Williams, and K. Flanigan. 1994. Effects of commercial vessel passage in narrow channels with and without ice cover. *Journal of Cold Regions Engineering* 8:47–64.

Lyon, J., D. Yuan, R. Lunetta, and C. Elvidge. 1998. A change detection experiment using vegetation indices. *Photogrammetric Engineering and Remote Sensing* 64:143–150.

MacArthur, R., and E. Wilson. 1967. *The theory of island biogeography.* Princeton, NJ: Princeton University Press.

Maehr, D., and J. Brady. 1984. Food habits of Florida black bears. *Journal of Wildlife Management* 48:230–235.

Magurran, A. 1988. *Ecological diversity and its measurement.* Princeton, NJ: Princeton University Press.

Maidment, D. 2002. *Arc hydro: GIS for water resources.* Redlands, CA: ESRI Press.

Maidment, D., and D. Djokic. 2000. *Hydrologic and hydraulic modeling support with GIS.* Redlands, CA: ESRI Press.

Maltby, E., and T. Barker. 2009. *The wetlands handbook.* New York, NY: Wiley-Blackwell.

Marks, M., B. Lapin, and J. Randall. 1994. Phragmites australis (P. communis): Threats, management, and monitoring. *Natural Areas Journal* 14:285–294.

Martz, L., and J. Garbrecht. 2003. Automated extraction of drainage network and watershed data from digital elevation models. *Water Resources Bulletin* 29:901–908.

Matula, G. 1974. Behavioral and physiological characteristics of black bear in northeastern Pennsylvania. Master's thesis, Pennsylvania State University.

May, C., R. Horner, J. Karr, B. Mar and E. Welch. 1997. Effects of urbanization on small streams in the Puget Sound Lowland Ecoregion. *Watershed Protection Techniques* 2:483–493.

May, D., J. Wang, J. Kovacs, and M. Muter. 2002. *Mapping wetland extent using IKONOS satellite imagery of the O'Donnell Point region, Georgian Bay, Ontario.* London, Ontario: Department of Geography, the University of Western Ontario.

McArthur, K. 1981. Factors contributing to effectiveness of black bear transplants. *Journal of Wildlife Management* 45:102–110.

MacAuthur, R.H., and Wilson, E.O. 1967. *The Theory of Island Biogeography.* Princeton, NJ: Princeton University Press.

McCauley, L., and D. Jenkins. 2005. GIS-based estimates of former and current depressional wetlands in an agricultural landscape. *Ecological Applications* 15:1199–1208.

McDonnell, M. 1984. Interactions between landscape elements: Dispersal of bird disseminated plants in post-agricultural landscapes. In J. Brandt, and R. Agger, eds., *Methodology in landscape ecological research and planning.* Roskilds, Denmark: International Association of Landscape Ecologists.

McDonnell, M., and E. Stiles. 1983. The structural complexity of old field vegetation and the recruitment of bird-dispersed plant species. *Oecologia* 56:109–116.

McGarigal, K. 2002. Landscape pattern metrics. In A. El-Shaarawi and W. Piegorsch, eds., *Encyclopedia of environmetrics,* vol. 2. Sussex, UK: J. Wiley & Sons.

McIntyre, N., and J. Wiens. 1999a. How does habitat patch size affect animal movement? An experiment darkling beetles. *Ecology* 80:2262–2270.

McIntyre, N., and J. Wiens. 1999b. Interactions between habitat abundance and configuration: Experiment validation of some predictions from percolation theory. *Oikos* 86:129–137.

MEA (Millennium Ecosystem Assessment), 2005. Ecosystems and Human Well-being: Synthesis. Island Press, Washington, DC. 137pp.

Mealey, S. 1975. The natural food habits of free ranging grizzly bears in Yellowstone National Park, 1973–1974. Master's thesis, Montana State University.

Mehaffey, M., M. Nash, T. Wade, D. Ebert, K. Jones, and A. Rager. 2005. Linking land cover and water quality in New York City's water supply watersheds. *Environmental Monitoring and Assessment* 107:29–44.

Middleton, B. 1999. *Wetland restoration, flood pulsing and disturbance dynamics.* New York, NY: J. Wiley & Sons.

Miller, R.B., and C. Small. 2003. Cities from space: Potential applications of remote sensing in urban environmental research and policy. *Environ. Sci. Policy* 6, 129–137.

Miller, W., and F. Egler. 1950. Vegetation of the Wequetequock-Paw-Catuck tidal marshes, Connecticut. *Ecological Monographs* 20:147–171.

Millman, A. 1999. *Mathematical principles of remote sensing: Making inferences from noisy data.* Boca Raton, FL: CRC Press.

Milly, P., J. Betancourt, M. Falkenmark, R. Hirsch, Z. Kundzewicz, D. Lettenmaier, and R. Stouffer. 2008. Stationarity is dead: Whither water management? *Science* 319: 573–574.

Milly, P., R. Wetherald, K. Dunne, and T. Delworth. 2002. Increasing risk of great floods in a changing climate. *Nature* 415: 514–517.

Mitsch, W., J. Day, J. Gilliam, P. Groffman, D. Hey, G. Randall, and N. Wang. 2001. Reducing nitrogen loading to the Gulf of Mexico from the Mississippi river basin: Strategies to counter a persistent ecological problem. *Bioscience* 51: 373–388.

Mitsch, W., and J. Gosselink. 1993. *Wetlands.* Van Nostrand Reinhold, New York, NY, 722 pp.

Mitsch, W., and J. Gosselink. 2000. *Wetlands.* 3rd ed. New York, NY: J. Wiley & Sons.

Mitsch, W., and J. Gosselink. 2007. *Wetlands.* 4th ed. New York, NY: Van Nostrand Reinhold.

Mitsch, W., J. Gosselink, L. Zhang, and C. Anderson. 2009. *Wetland ecosystems.* New York, NY: J. Wiley & Sons.

Mitsch, W., and S. Jorgensen. 2004. *Ecological engineering and ecosystem restoration.* Hoboken, NJ: J. Wiley & Sons.

Mizgalewicz, P., W. White, D. Maidment, and M. Ridd. 2003. GIS modeling and visualization of the water balance during the 1993 Midwest flood. In J. Lyon, *Geographic information system applications for watershed and water resources management.* London, UK: Taylor & Francis.

Moik, J., 1980. *Digital processing of remotely sensed images.* NASA SP-431. Washington, DC: National Aeronautics and Space Administration.

Molina, M.J., and F.S. Rowland. 1974. Stratospheric sink for chlorofluoromethands: Chlorine atom-catalysed destruction of ozone. *Nature* 249, 810–812.

Moller, T., and C. Rordam. 1985. Species numbers of vascular plants in relation to area, isolation and age of ponds in Denmark. *Oikos* 45:8–16.

Mueller-Dombois, D., and H. Ellenberg. 1974. *Aims and methods of vegetation ecology.* New York, NY: J. Wiley & Sons.

Murkin, H., and X. Kale. 1986. Relationships between waterfowl and macroinvertebrate densities in a northern prairie marsh. *Journal of Wildlife Management* 50:212–217.

Na, X., S. Zhang, X. Li, H. Yu, and C. Liu. 2010. Improved land cover mapping using random forests combined with Landsat Thematic Mapper imagery and ancillary geographic data. *Photogrammetric Engineering and Remote Sensing* 76:833–840.

Nagasaka, A., and F., Nakamura. 1999. The influences of land-use changes on hydrology and riparian environment in a northern Japanese landscape. *Landscape Ecology* 14:543–556.

NASA. *Ozone Hole Watch*; Available online: http://ozonewatch.gsfc.nasa.gov/ (accessed on 1 March 2011).

Nash, M., D. Chaloud, and R. Lopez. 2005. *Multivariate analyses (canonical correlation analysis and partial least Square, PLS) to model and assess the association of landscape metrics to surface water chemical and biological properties using Savannah River basin data.* EPA/600/X-05/004. Washington, DC: U.S. Environmental Protection Agency.

National Research Council. 1995. *Wetlands: Characteristics and boundaries.* Washington, DC: National Academies Press.

National Research Council. 2001. *Compensating for wetland losses under the Clean Water Act.* Washington, DC: National Academies Press.

Natural Resources Conservation Service. 1997. Hydrology tools for wetland determination. In *Engineering field handbook.* Washington, DC: U.S. Department of Agriculture.

Natural Resources Conservation Service. 2000. *1997 National Resources Inventory wetland data.* Beltsville, MD: Natural Resources Conservation Service.

National Wetlands Inventory. 2012. NOAA Coastal Services Center digital coast website and U.S. Fish and Wildlife Service, accessed June 3, 2012, at http://www.csc.noaa.gov/digitalcoast/data/nwi or http://www.fws.gov/wetlands/.

National Wetland Plants List. 2012. http://geo.usace.army.mil/wetland_plants/index.html. Accessed June 3, 2012.

The Nature Conservancy. 1992. *The forested wetlands of the Mississippi River: An ecosystem in crisis.* Baton Rouge, LA: The Nature Conservancy.

Nestler, J., and K. Long. 1994. *Cumulative impact analysis of wetlands using hydrologic indices.* Technical Report WRP-SM-3. Vicksburg, MS: U.S. Army Engineer Waterways Experiment Station.

Nichols, J., K. Reinecke, and J. Hines. 1983. Factors affecting the distribution of mallards wintering in the Mississippi alluvial valley. *Auk* 100:932–946.

Niedzwiedz, W., and L. Ganske. 1991. Assessing lakeshore permit compliance using low altitude oblique 35-mm aerial photography. *Photogrammetric Engineering and Remote Sensing* 57:511–518.

Niering, W., and R. Warren. 1980. Vegetation patterns and processes in New England salt marshes. *BioScience* 30:301–307.

Nip-van der Voort, J., R. Hengeveld, and J. Haeck. 1979. Immigration rates of plant succession in three Dutch polders. *Journal of Biogeography* 6: 301–308.

Noble, I. 1989. Attributes of invaders and the invading process: Terrestrial and vascular plants. In J. Drake, H. Mooney, F. di Castri, R. Groves, F. Kruger, M. Rejmanek, and M. Williamson, eds., *Biological invasions: A global perspective.* Chichester, UK: J. Wiley & Sons.

Odum, E. 1985. Trends expected in stressed ecosystems. *Bioscience* 35:419–422.

Office of Science and Technology Policy. 2006. *Improved observations for disaster reduction: Near-term opportunity plan.* Washington, DC: Subcommittee on Disaster Reduction, Executive Offices of the President.

Office of Science and Technology Policy. 2010. *Achieving and sustaining Earth observations*: A preliminary plan based on a strategic assessment by the U.S. Group on Earth Observations. Washington DC: The White House.

Ogutu, Z. 1996. Multivariate analysis of plant communities in the Narok district, Kenya: The influence of environmental factors and human disturbance. *Vegetatio* 126:181–189.

O'Hara, C., T. Cary, and K. Schuckman. 2010. Integrated technologies for orthophoto accuracy verification and review. *Photogrammetric Engineering and Remote Sensing* 76:1097–1103.

Omernik, J. 1987. Ecoregions of the conterminous United States. *Annals of the Association of American Geographers* 77:188–125.

Omernik, J. 2003. The misuse of hydrologic unit maps for extrapolation, reporting, and ecosystem management. *Journal of the American Water Resources Association* 39:563–573.

O'Neill, R. 2001. Predicting nutrient and sediment loadings to streams from landscape metrics: A multiple watershed study from the United States Mid-Atlantic Region, *Landscape Ecology* 16:301–312.

O'Neill, R., C. Hunsaker, and D. Levine. 1992. Monitoring challenges and innovative ideas. In D. McKenzie, D., Hyatt, and V. McDonald, eds., *Ecological indicators*. London, UK: Elsevier Applied Science.

Opdam, P. 1990. Understanding the ecology of populations in fragmented landscapes. Trondheim, Norway. *Transactions of the 19th IUGB Congress*. Trondheim, Norway: Norwegian Institute for Nature Research.

Opdam, P., R. Apeldoorn, A. Schotman, and J. Kalkhoven. 1993. Population responses to landscape fragmentation. In C. Vos, and P. Opdam, eds. *Landscape ecology of a stressed environment*. London, UK: Chapman & Hall.

Ozesmi, S., and M. Bauer. 2002. Satellite remote sensing of wetlands. *Wetlands Ecology and Management* 10: 381–402.

Ozoga, J. J., and L. J. Verme. 1982. Predation by black bears on newborn white-tailed deer. *Journal of Mammology*. 63:695–696.

Pantaleoni, E., R. Wynne, J. Galbraith, and J. Campbell. 2009. Mapping wetlands using ASTER data: A comparison between classification trees and logistic regression. *International Journal of Remote Sensing* 30: 3423–3440.

Parry, M., O. Canziani, J. Palutikof, P. Van der Linden, and C. Hanson. 2007. Climate change 2007: Impacts, adaption and vulnerability. Contribution of Working Group II to the *Fourth Assessment Report of the Intergovernmental Panel on Climate Change (IPCC)*. New York, NY. Cambridge University Press

Pelton, M. 1982. Black bear. In J. Chapman and G. Feldhamer, eds., *Wild mammals of North America: biology, management, and economics*. Baltimore, MD: Johns Hopkins University Press.

Peterjohn, W., and D. Corel. 1984. Nutrient dynamics in an agricultural watershed: Observations on the role of a riparian forest. *Ecology* 65:1466–1475.

Peterson, S., L. Carpenter, G. Guntenspergen, and L. M. Cowardin. 1996. *Pilot test of wetland condition indicators in the Prairie Pothole region of the United States*. EPA/620/R-97/002. Corvalis, OR: U.S. Environmental Protection Agency.

Philipson, W.R. (ed.). 1997. Manual of Photographic Interpretation. American Society for Photogrammetry and Remote Sensing. Bethesda, MD, USA.

Pickett, S., and J. Thompson. 1978. Patch dynamics and the design of natural reserves. *Biological Conservation* 13:27–37.

Place, J. 1985. Mapping of forested wetland: Use of Seasat radar images to complement conventional sources. *The Professional Geographer* 37:463–469.

Poiani, K., and P. Dixon. 1995. Seed banks of Carolina bays: Potential contributions from surrounding landscape vegetation. *American Midland Naturalist* 134:140–154.

Pollard, T., I. Eden, J. Mundy, and D. Cooper. 2010. A volumetric approach to change detection in satellite images. *Photogrammetric Engineering and Remote Sensing* 76:817–831.

Prigent, C., E. Matthews, F. Aires, and W. Rossow. 2001. Remote sensing of global wetland dynamics with multiple satellite data sets. *Geophysical Research Letters* 28:4631–4634.

Raabe, E., and R. Stumpf, 1995. Monitoring tidal marshes of Florida's Big Bend. Proceedings of the Third Thematic Conference on Remote Sensing of Marine and Coastal Environments, ERIM, Ann Arbor, MI, Vol. 2, pp. 483–494.

Radwell, A. J., and T. J. Kwak. 2005. Assessing ecological integrity of Ozark rivers to determine suitability for protective status. *Environmental Management* 35:799–810.

Ramsar. 2002. *The Ramsar list of wetlands of international importance.* Convention on Wetlands, Article 2.1, 1989 amendment. Ramsar, Iran. Ramsar Convention.

Ramsey, E. 1995. Monitoring flooding in coastal wetlands by using radar imagery and ground-based measurements. *International Journal of Remote Sensing* 16:2495–2502.

Ramsey, E. 1998. Radar remote sensing of wetlands. In R. Lunetta and C. Elvidge, *Remote sensing change detection.* Chelsea, MI, and Boca Raton, FL: Ann Arbor Press/CRC Lewis Publishers.

Ramsey, E. 2005. Remote sensing of coastal environments. In M. Schwartz, ed., *Encyclopedia of coastal science* (Encyclopedia of Earth Sciences Series). Dordrecht, Netherlands: Springer.

Ramsey, E., D. Chappell, D. Jacobs, S. Sapkota, and D. Baldwin. 1998. Resource management of forested wetlands: Hurricane impact and recovery mapped by combining Landsat TM and NOAA AVHRR data. *Photogrammetric Engineering and Remote Sensing* 64:733–738.

Ramsey, E., and J. Jensen. 1990. The derivation of water volume reflectances from airborne MSS data using in situ water volume reflectances, and a combined optimization technique and radiation transfer model. *International Journal of Remote Sensing* 11:979–998.

Ramsey, E., and J. Jensen. 1995. Modelling mangrove canopy reflectance using a light interaction model and an optimization technique. In J. Lyon and J. McCarthy, *Wetland and environmental applications of GIS.* Boca Raton, FL: Lewis Publishers.

Ramsey, E., and J. Jensen. 1996. Remote sensing of mangrove wetlands: Relating canopy spectra to site-specific data. *Photogrammetric Engineering and Remote Sensing* 62:939–948.

Ramsey, E., and S. Laine. 1997. Comparison of Landsat Thematic Mapper and high resolution photography to identify change in complex coastal marshes. *Journal of Coastal Research* 13:281–292.

Ramsey, E., S. Laine, D. Werle, B. Tittley, and D. Lapp. 1994. Monitoring Hurricane Andrew damage and recovery of the coastal Louisiana marsh using satellite remote sensing data. In P. Wells and P. Ricketts, eds., *Proceedings of the Coastal Zone.* Hallifax, Nova Scotia: Canada.

Ramsey, E., Z. Lu, Y. Suzuoki, A. Rangoonwala, and D. Werle. 2011. Monitoring duration and extent of storm surge flooding along the Louisiana coast with Envisat ASAR data. *IEEE Journal of Selected Topics on Applied Earth Observations and Remote Sensing* 4:387–399.

Ramsey, E., G. Nelson, F. Baarnes, R. Spell. 2004. Light attenuation profiling as an indicator of structural changes in coastal marshes. In R. Lunetta J. and J. Lyon, eds., *Remote sensing and GIS accuracy assessment.* New York, NY: CRC Press.

Ramsey, E., G. Nelson, and S. Sapkota. 1998. Classifying coastal resources by integrating optical and radar imagery and color infrared photography. *Mangroves and Salt Marshes* 2:109–119.

Ramsey E., G. Nelson, and S. Sapkota. 2001. Coastal change analysis program implemented in Louisiana. *Journal of Coastal Research* 17:55–71.

Ramsey, E., G. Nelson, S. Sapkota, S. Laine, J. Verdi, and S. Krasznay. 1999. Using multiple polarization L-band radar to monitor marsh burn recovery. *IEEE Transactions Geoscience and Remote Sensing* 37:635–639.

Ramsey, E., G. Nelson, S. Sapkota, M. Strong, W. Phillips, and K. Schmersahl. 2001a. *Landcover classification of Lake Meredith National Recreation Area.* USGS/BRD/BSR—2001-0003. U.S. Geological Survey.

Ramsey, E. and A. Rangoonwala. 2006. Site-specific canopy reflectance related to marsh dieback onset and progression in coastal Louisiana. *Photogrammetric Engineering and Remote Sensing* 72:641–652.

Ramsey, E., A. Rangoonwala, B. Middleton, and Z. Lu. 2009. Satellite optical and radar image data of forested wetland impact on and short-term recovery from Hurricane Katrina in the lower Pearl River flood plain of Louisiana, USA. *Wetlands* 29:66–79.

Ramsey, E., A. Rangoonwala, Y. Suzuoki, and C. Jones. 2011. Oil detection in a coastal marsh with polarimetric synthetic aperture radar (SAR). *Journal of Remote Sensing* 3:2630–2662.

Ramsey, E., D. Werle, Y. Suzuoki, A. Rangoonwala, and Z. Lu. 2012. Limitations and potential of satellite imagery to monitor environmental response to coastal flooding. *Journal of Coastal Research* 28:457–476.

Ramsey, E., D. Werle, Z. Lu, A. Rangoonwala, and Y. Suzuoki. 2009. A case of timely satellite image acquisitions in support of coastal emergency environmental response management. *Journal of Coastal Research* 25:1168–1172.

Reddy, K., and R. DeLaune. 2008. *Biogeochemistry of wetlands: Science and applications.* Boca Raton, FL: CRC Press.

Reed, P. 1988. *National list of plant species that occur in wetlands: National summary.* Washington, DC: U.S. Department of Interior.

Reid, W.V., D. Chen, L. Goldfarb, H. Hackmann, Y.T. Lee, K. Mokhele, E. Ostrom, K. Raivio, J. Rockström, H.J. Schellnhuber, and A. Whyte. 2010. Earth system science for global sustainability: Grand challenges. *Science* 330, 916–917.

Remillard, M., and R. Welch. 1993. GIS technologies for aquatic macrophyte studies: Modeling applications. *Landscape Ecology* 8:163–175.

Research Systems, Inc. (RSI). 2001. *ENVI User's Guide*, ENVI v. 3.5. Boulder, CO: RSI.

Reynolds, D.G., and J.J. Beecham. 1980. Home range activities and reproduction of black bears in west-central Idaho. *Bear Biological Association Conference Series.* 3:181–190.

Rheinhardt, R., M. Rheinhardt, and M. Brinson. 2002. *A regional guidebook for applying the hydrogeomorphic approach to assessing wetland functions of wet pine flats on mineral soils in the Atlantic and Gulf coastal plains.* ERDC/EL TR-02-9. Vicksburg, MS: U.S. Army Engineer Research and Development Center.

Richardson, J., and M. Vepraskas. 2007. *Wetland soils: Genesis, hydrology, landscapes, and classification.* Boca Raton, FL: CRC Press.

Ridley, H.N. 1930. The Dispersal of Plants Throughout the World. L. Reeve and Co. LTD, Ashford, Kent, UK.

Riitters, K., and J. Coulston. 2005. Hot spots of perforated forest in the Eastern United States. *Environmental Management* 35:483–492.

Riitters, K., R. O'Neill, C. Hunsaker, J. Wickham, D. Yankee, S. Timmins, K. Jones, and B. Jackson. 1995. A factor analysis of landscape pattern and structure metrics. *Landscape Ecology* 10:23-39.

Riitters, K., J. Wickham, R. O'Neill, K. Jones, and E. Smith. 2000. Global-scale patterns of forest fragmentation. *Conservation Ecology* 4:3.

Robinson, G., R. Holt, M. Gaines, S. Hambur, M. Johnson, H. Fitch, and E. Martinko. 1992. Diverse and contrasting effects of habitat fragmentation. *Science* 257:524–526.

Rodrigues-Galiano, V., B. Ghimire, E. Pardo_Iguzquiza, M. Chica-Olmo, R. Congalton. 2012. Incorporating the downscaled Landsat TM thermal band in land-cover classification using random forest. *Photogrammetric Engineering and Remote Sensing* 78:129–137.

Rogers, L. 1976. Effects of mast and berry crop failures on survival, growth, and reproductive success of black bears. *Transactions of the North American Wildlife and Natural Resources Conference* 41:432–438.

Rogers, L. 1987. Effects of food supply and kinship on social behavior, movements, and population growth of black bears in northeastern Minnesota. *Wildlife Monograph* 97:1–72.

Rogers, L. 1992. *Watchable wildlife: The black bear*. Minneapolis, MN: U.S. Forest Service, North Central Forest Experiment Station.

Rogers, L., and A. Allen. 1987. *Habitat suitability index models: Black bear, upper Great Lakes region*. Biological Report 82 (10.144). Washington, DC: Department of Interior, U.S. Fish and Wildlife Service.

Roller, N. 1977. *Remote sensing of wetlands*. Technical Report No. 193400-14-T. Ann Arbor, MI: Environmental Research Institute of Michigan.

Roth, N., J. Allan, and D. Erickson. 1996. Landscape influences on stream biotic integrity assessed at multiple spatial scales. *Landscape Ecology* 11:141–156.

Rouse, J., R. Haas, J. Schell, and D. Deering. 1973. Monitoring vegetation systems in the Great Plains with ERTS. Paper presented at the Third ERTS Symposium, NASA SP-351.

Rowan, W. 1928. Bears and bird eggs. *Condor* 30:246.

Rundquist, D., S. Narumalani, and R. Narayanan. 2001. A review of wetlands remote sensing and defining new considerations. *Remote Sensing Reviews* 20:207–226.

Running, S.W., R.R. Nemani, F.A. Heinsch, M. Zhao, M. Reeves, and H. Hasimoto. 2004. A continuous satellite-derived measure of global terrestrial primary production. *BioScience* 54, 547–560.

Sader, S., D. Ahl, and W. Liou. 1995. Accuracy of Landsat-TM and GIS rule-based methods for forest wetland classification in Maine. *Remote Sensing of Environment* 53:133–144.

Salvesen, D. 1990. *Wetlands: Mitigating and regulating development impacts*. Washington, DC: The Urban Land Institute.

SAS Institute. 1998. *Version 9 user's guide*. Cary, NC: SAS Institute Inc.

Schaal, G. 1995. *Methods used in the Ohio Wetland Inventory*. Columbus, OH: Ohio Department of Natural Resources.

Schlesinger, W. 1997. *Biogeochemistry: An analysis of global change*. San Diego, CA: Academic Press.

Schott, J. 2007. *Remote sensing: The image chain approach*. New York, NY: Oxford University Press.

Schowengerdt, R. 2007. Remote sensing: Models and methods for image processing. Burlington, MA: Academic Press.

Schroeder, R. 1996a. *Wildlife community habitat evaluation: A model for deciduous palustrine forested wetlands in Maryland*. Technical Report WRP-DE-14. Vicksburg, MS: U.S. Army Engineer Waterways Experiment Station.

Schroeder, R. 1996b. *Wildlife community habitat evaluation using a modified species-area relationship*. Technical Report WRP-DE-12. Vicksburg, MS: U.S. Army Engineer Waterways Experiment Station.

Scieszka, M. 1990. *The digital wetlands data base for the U.S. Great Lakes shoreline*. Federal Coastal Wetland Mapping Programs, Biology Report 90 (18). Washington, DC: Department of Interior, U.S. Fish and Wildlife Service.

Scott, J., F. Davis, B. Csulti, R. Noss, B. Butterfield, C. Groves, H. Anderson, S. Caicco, F. D'Erchia, T. Edwards, J. Ulliman, and R. Wright. 1993. Gap analysis: a geographic approach to protection of biological diversity. *Wildlife Monographs* 123:1–41.

Sculthorpe, C. 1967. *The biology of aquatic vascular plants*. London, UK: Arnold.

Sen, S., C. Zipper, R. Wynne, and P. Donovan. 2012. Identifying revegetated mines as disturbance/recovery trajectories using an interannual Landsat chronosequence. *Photogrammetric Engineering and Remote Sensing* 78:223–235.

Shafer, D., B. Herczeg, D. Moulton, A. Sipocz, K. Jaynes, L. Rozas, C. Onuf, and W. Miller. 2002. *Regional guidebook for applying the hydrogeomorphic approach to assessing wetland functions of Northwest Gulf of Mexico tidal fringe wetlands*. ERDC/EL TR-02-5. Vicksburg, MS: U.S. Army Engineer Research and Development Center.

Shafer, D., J. Karazsia, L. Carrubba, and C. Martin. 2008. *Evaluation of regulatory guidelines to minimize impacts to seagrasses from single family residential dock structures in Florida and Puerto Rico*. ERDC/EL TR-08-41. Vicksburg, MS: U.S. Army Engineer Research and Development Center.

Shafer, D., T. Roberts, M. Peterson, and K. Schmid. 2007. *A regional guidebook for applying the hydrogeomorphic approach to assessing the functions of tidal fringe wetlands along the Mississippi and Alabama Gulf Coast*. ERDC/EL TR-07-2. Vicksburg, Miss.: U.S. Army Engineer Research and Development Center.

Shafer, D., and D. Yozzo. 1998. *National guidebook for application of hydrogeomorphic assessment of tidal fringe wetlands*. Technical Report WRP-DE-16. Vicksburg, MS: U.S. Army Engineer Waterways Experiment Station.

Shan, J., and C. Toth. 2008. *Topographic laser ranging and scanning: Principles and processing*. Boca Raton, FL: CRC Press.

Shaw, D., D. Field, T. Holm, M. Jennings, J. Sturdevant, G. Thelin, and L. Worthy. 1993. An innovative partnership for national environmental assessment. 12th Pecora Symposium, Sioux Falls, South Dakota.

Shaw, S., and C. Fredine. 1956. *Wetlands of the United States*. Circular 39. Washington, DC: Department of Interior, U.S. Fish and Wildlife Service.

Shuman, C., and R. Ambrose. 2003. A comparison of remote sensing and ground-based methods for monitoring wetland restoration success. *Restoration Ecology* 11:325–333.

Simberloff, D., and L. Abele. 1982. Refuge design and island biogeographic theory: Effects of fragmentation. *The American Naturalist* 120:41–50.

Simberloff, D., and E. Wilson. 1970. Experimental zoogeography of islands: A two-year record of colonization. *Ecology* 51:934–937.

Simpson, J., R. Boerner, M. DeMers, L. Berns, F. Artigas, and A. Silva. 1994. Forty-eight years of landscape change on two contiguous Ohio landscapes. *Landscape Ecology* 9:261–270.

Singh, A., 1989. Digital change detection techniques using remotely-sensed Data. *International Journal of Remote Sensing* 10:989–1003.

Sinha, P., L. Kumar, and N. Reid. 2012. Seasonal variation in land-cover classification accuracy in a diverse region. *Photogrammetric Engineering and Remote Sensing* 78:271–280.

Skaggs, R., D. Amatya, R. Evans, and J. Parsons. 1994. Characterization and evaluation of proposed hydrologic criteria for wetlands. *Journal of Soil and Water Conservation* 49:501–510.

Smith, K.G., R.S. Dzur, D.G. Catanzaro, M.E. Garner, W.F. Limp. 1998. The Arkansas GAP Analysis Project. Final Report [Available online at URL: http://web.cast.uark.edu/gap/].

Smith, R. 1996. Composition, structure, and distribution of woody vegetation on the Cache River floodplain, Arkansas. *Wetlands* 16:264–278.

Smith, R., and J. Wakeley. 2001. Developing assessment models. In *Hydrogeomorphic approach to assessing wetland functions: Guidelines for developing regional guidebooks*. ERDC/EL TR-01-30. Vicksburg, MS: U.S. Army Engineer Research and Development Center.

Smith, T. 1985. Ecology of black bears in a bottomland hardwood forest in Arkansas. Doctoral dissertation, University of Tennessee.

Soils Conservation Service. 1992. State soils geographic database (STATSGO) user's guide. Soil Conservation Service Publication, No. 1492.

Soule, M., B. Wilcox, and C. Holtby. 1979. Benign neglect: A model of faunal collapse in game reserves of East Africa. *Biological Conservation* 15:259–272.

Sprecher, S., and C. Schneider. 2000. *Wetlands management handbook*. ERDC/EL SR-00-16. Vicksburg, MS: U.S. Army Engineer Research and Development Center.

Sprecher, S., and A. Warne. 2000. *Accessing and using meteorological data to evaluate wetland hydrology*. ERDC/EL TR-WRAP-00-01. Vicksburg, MS: U.S. Army Engineer Research and Development Center.

State of the Lakes Ecosystem Conference. 2000. Selection of indicators for Great Lakes basin ecosystem health. U.S. Environmental Protection Agency.

Stevens, D., and S. Jensen. 2007. Sample design, execution, and analysis for wetland assessment. *Wetlands* 27:515–523.

Stiling, P. 1996. *Ecology: Theories and applications*. 2nd ed. Upper Saddle River, NJ: Prentice Hall.

Stoddart, D. 1965. The shape of atolls. *Marine Geology* 3:369–383.

Stone, R. 2010.Earth-observation summit endorses global data sharing. *Science* 330, 902.

Stritholt, J., and R. Boerner. 1995. Applying biodiversity gap analysis in a regional nature reserve design for the edge of Appalachia, Ohio (U.S.A.). *Conservation Biology* 9: 1492–1505.

Stuckey, R. 1989. Western Lake Erie aquatic and wetland vascular-plant flora: Its origin and change. In K. Krieger, ed., *Lake Erie estuarine systems: Issues, resources, status, and management*. Washington, DC: Department of Commerce, National Oceanographic and Atmospheric Administration.

Stutheit, R., M. Gilbert, P. Whited, and K. Lawrence. 2004. *A regional guidebook for applying the hydrogeomorphic approach to assessing wetland functions of rainwater basin depressional wetlands in Nebraska*. ERDC/EL TR-04-4. Vicksburg, MS: U.S. Army Engineer Research and Development Center.

Sun, D., Y. Yu, R. Zhang, S. Li, and M. Goldberg. 2012. Towards operational automatic flood detection using EOS/MODIS data. *Photogrammetric Engineering and Remote Sensing* 78:637–646.

Svejkovsky, J., and J. Muskat. 2006. *Real-time detection of oil slick thickness patterns with a portable multispectral sensor.* Final report. Herndon, VA: U.S. Minerals Management Service.

Tietje, W., B. Pelchat, and R. Ruff. 1986. Cannibalism of denned black bears. *Journal of Mammology* 67:762–766.

Tiner, R. 1999. *Wetland indicators: A guide to wetland identification, delineation.* Boca Raton, FL: CRC Press.

Tiner, R. 2003a. Estimated extent of geographically isolated wetlands in selected areas of the U.S. *Wetlands* 23:636–652.

Tiner, R. 2003b. Geographically isolated wetlands of the United States. *Wetlands* 23:494–516.

Tiner, R., H. Bergquist, G. DeAlessio, and M. Starr. 2002. *Geographically isolated wetlands: A preliminary assessment of their characteristics and status in selected areas of the U.S.* Hadley, MA: Department of the Interior, U.S. Fish and Wildlife Service.

Thenkabail, P. 2006. Limpopo river basin in Southern Africa. IWMIs Challenge Programme River Basins. Available at http://www.wetlands.iwmids.org. Accessed December 12, 2012.

Thenkabail, P., C. Biradar, H. Turral, P. Noojipady, Y. Li, J. Vithanage, V. Dheeravath, M. Velpuri, M. Schull, X. L. Cai, and R. Dutta. 2006. *An irrigated area map of the world (1999) derived from remote sensing.* IWMI Research Report. Colombo, Sri Lanka: International Water Management Institute.

Thenkabail, P., J. Lyon, and A. Huerte. 2012. *Hyperspectral remote sensing of vegetation.* Boca Raton, FL: CRC Press.

Thenkabail, P., J. Lyon, H. Turral, and C. Biradar. 2009. *Remote sensing of global croplands for food security.* Boca Raton, FL: CRC Press.

Thenkabail, P., C. Nolte, and J. Lyon. 2000. Remote sensing and GIS modeling for selection of benchmark research area in the inland valley agroecosystems of West and Central Africa. *Photogrammetric Engineering and Remote Sensing* 66:755–768.

Thenkabail, P., M. Schull, and H. Turral. 2005. Ganges and Indus river basin land use/land cover (LULC) and irrigated area mapping using continuous streams of MODIS data. *Remote Sensing Environment* 95:317–341.

Thompson, G., and R. Gauthier. 1990. Development of a GIS for the U.S. Great Lakes shoreline. Unpublished manuscript, Detroit District, U.S. Army Corps of Engineers.

Torbick, N. 2004. The utilization of remote sensing and geographic information systems (GIS) for the development of a wetlands classification and inventory for the Lower Maumee River watershed, Lucas County, Ohio. Master's thesis, University of Toledo.

Touzi, R., A. Deschamps, and G. Rother. 2007. Wetland characterization using polarimetric Radarsat-2 capability. *Canadian Journal of Remote Sensing* 33:S56–S67.

Turner, M., R. Garner, and R. O'Neill. 2001. *Landscape ecology in theory and practice.* New York, NY: Springer-Verlag.

Turner, R. E., and N. N. Rabalais. 2004. Suspended sediment, C, N, P, and Si yields from the Mississippi River Basin. *Hydrobiologia* 511:79–89.

Twedt, D., and C. Loesch. 1999. Forest area and distribution in the Mississippi alluvial valley: Implications from breeding bird conservation. *Journal of Biogeography* 26:1215–1224.

Unwin, D. 1981. *Introductory spatial analysis.* London, UK: Methuen.

Uranowski, C., Z. Lin, M. DelCharco, C. Huegel, J. Garcia, I. Bartsch, M. Flannery, S. Miller, J. Bacheler, and W. Ainslie. 2003. *A regional guidebook for applying the hydrogeomorphic approach to assessing wetland functions of low-gradient, blackwater riverine wetlands in peninsular Florida.* ERDC/EL TR-03-3. Vicksburg, MS: U.S. Army Engineer Research and Development Center.

U.S. Army Corps of Engineers. 1987. *Corps of Engineers wetlands delineation manual.* Technical Report Y-87-1. Vicksburg, MS: Environmental Laboratory, U.S. Army Engineer Waterways Experiment Station.

U.S. Army Corps of Engineers. 1988. *Draft environmental impact statement, supplement II to the final environmental impact statement, operations, maintenance, and minor improvements of the federal facilities at Sault Ste. Marie, Michigan, July, 1977.* Detroit, MI: U.S. Army Corps of Engineers District.

U.S. Army Corps of Engineers. 1993. *Photogrammetric mapping.* Engineering manual. Washington, DC.

U.S. Army Corps of Engineers. 2009. *Interim regional supplement to the Corps of Engineers wetland delineation manual: Northcentral and northeast region.* Wetlands Regulatory Assistance Program, ERDC/EL TR-09-19. Vicksburg, MS: U.S. Army Engineer Research and Development Center.

U.S. Army Corps of Engineers. 2010a. *Regional supplement to the Corps of Engineers wetland delineation manual: Great Plains region (v. 2.0).* ERDC/EL TR-10-1. Vicksburg, MS: U.S. Army Engineer Research and Development Center.

U.S. Army Corps of Engineers. 2010b. *Regional supplement to the Corps of Engineers wetland delineation manual: Western mountains, valleys, and coast region (v. 2.0).* ERDC/EL TR-10-3. Vicksburg, MS: U.S. Army Engineer Research and Development Center.

U.S. Department of Agriculture. 1962. *Soil survey manual.* Washington, DC: Soil Conservation Service.

U.S. Department of Agriculture. 1975. *Soil taxonomy, a basic system of soil classification for making and interpreting soil surveys.* Washington, DC: U.S. Soil Conservation Service.

U.S. Department of Agriculture. 1988. *National Food Security Act manual.* Washington, DC: U.S. Department of Agriculture.

U.S. Department of Agriculture. 1991. *Hydric soils of the United States.* Miscellaneous Publication 1491. Washington, DC: Soil Conservation Service.

U.S. Department of Agriculture. 1996. *Field indicators of hydric soils in the United States.* Washington, DC: U.S. Department of Agriculture, Natural Resources Conservation Service.

U.S. Department of Agriculture. 2001. *Summary report: 1997 National Resources Inventory (revised December 2001).* Washington, DC: U.S. Department of Agriculture, Natural Resource Conservation Service.

U.S. Department of the Interior. 2003. *Riparian area management: Riparian-wetland soils.* Tech. Rep. 1737-19, BLM/ST/ST-03/001+1737. Denver, CO: Bureau of Land Management.

U.S. Environmental Protection Agency. 1991. Federal manual for identifying and delineating jurisdictional wetlands. *Federal Register*, August 14.

U.S. Environmental Protection Agency. 1995. *Ecological restoration: A tool to manage stream quality.* EPA/841/F-95/007. Washington, DC: United States Environmental Protection Agency.

U.S. Environmental Protection Agency. 1997. *Monitoring water quality: Volunteer stream monitoring—A methods manual*. EPA/841/B-97/003. Washington, DC: U.S. Environmental Protection Agency.

U.S. Environmental Protection Agency. 1999. http://eh2o.saic.com/Documentation/ ICWater_AWR_Impact_Journal.pdf. Accessed January 10, 2013.

U.S. Environmental Protection Agency. 2001a. *Landscape analysis and assessment— Overview*. Las Vegas, NV: U.S. Environmental Protection Agency.

U.S. Environmental Protection Agency. 2001b. *National coastal condition report*. EPA-620/R-01/005. Washington, DC: U.S. Environmental Protection Agency, Office of Research and Development, Office of Water.

U.S. Environmental Protection Agency. 2008. *Methods for evaluating wetland condition: Wetland hydrology*. EPA-822-R-08-024. Washington, DC: U.S. Environmental Protection Agency, Office of Water.

U.S. Fish and Wildlife Service. 2004. *Technical procedures for wetlands status and trends*. Operational Version. Arlington, VA: Branch of Habitat Assessment.

U.S. Geological Survey. 1999. *Effects of animal feeding operations on water resources and the environment*. Open-File Report 00-204. Technical Meeting Proceedings, Fort Collins, CO: U.S. Geological Survey.

U.S. Geological Survey. 2003. *United States Geologic Survey: Effects of urban development on floods*. http://pubs.usgs.gov/fs/fs07603/pdf/fs07603.pdf. Accessed July 2012.

U.S. Geological Survey. 2006. *Flood hazards—A national threat*. http://pubs.usgs.gov/ fs/2006/3026/2006-3026.pdf. Accessed July 2012.

Van der Valk, A. 1981. Succession in wetlands: A Gleasonian approach. *Ecology* 62:688–696.

Van der Valk, A., and C. Davis. 1980. The impact of a natural drawdown on the growth of four emergent species in a prairie glacial marsh. *Aquatic Botany* 9:301–322.

Van Derventer, A. 1992. Evaluating the usefulness of Landsat Thematic Mapper to determine soil properties, management practices, and soil water content. Doctoral dissertation, Ohio State University .

Van Genderen, J., and B. Lock. 1977. Testing land-use map accuracy. *Photogrammetric Engineering and Remote Sensing* 43:1135–1137.

Van Remortel, R., R. Maichle, D. Heggem, and A. Pitchford. 2005. *Automated GIS watershed analysis tools for RUSLE/SEDMOD soil erosion and sedimentation modeling*. EPA/600/X-05/007. Washington, DC: United States Environmental Protection Agency.

Van Remortel, R., R. Maichle, and R. Hickey. 2004. Computing the RUSLE LS factor through array-based slope length processing of digital elevation data using a C++ executable. *Computers and Geosciences* 30:1043–1053.

Van Sickle, J. 2008. *GPS for land surveyors*. 3rd ed. Boca Raton, FL: CRC Press.

Vannote, R., G. Minshall, K. Cummins, J. Sedell, and C. Cushing. 1980. The river continuum concept. *Canadian Journal of Fish and Aquatic Science* 37:130–137.

Velpuri, N., P. Thenkabail, M. Gumma, C. Biradar, V. Dheeravath, P. Noojipady, and L. Yuanjie. 2009. Influence of resolution inn irrigated area mapping and area estimation. *Photogrammetric Engineering and Remote Sensing* 75:1383–1395.

Vepraskas, M., X. He, D. Lindbo, and R. Skaggs. 2004. Calibrating hydric soil field indicators to long-term wetland hydrology. *Soil Science Society of America Journal* 68:1461–1469.

Verbyla, D., and T. Hammond. 1995. Conservative bias in classification accuracy assessment due to pixel-by pixel comparison of classified images with reference grids: *International Journal of Remote Sensing* 16: 581–587.

Vicente-Serrano, S., F. Perez-Cabello, and T. Lasanta. 2008. Assessment of radiometric correction techniques in analyzing vegetation variability and change using time series of Landsat images. *Remote Sensing of Environment* 112: 3916–3934.

Villeneuve, J. 2005. Delineating wetlands using geographic information system and remote sensing technologies. Master's thesis, Texas A&M University.

Vogelmann, J.E., S.M. Howard, L. Yang, C.R. Larson, B.K. Wylie, and J.N. Van Driel, 2001, Completion of the 1990's National Land Cover Data Set for the conterminous United States, Photogrammetric Engineering and Remote Sensing 67:650–662.

Voss, E.G. 1972. Part I, Gymnosperms and Monocots. Michigan Flora. Cranbrook Institute of Science and University of Michigan Herbarium, Ann Arbor, Michigan, USA.

Voss, E.G. 1985. Part II, Dicots, Saururaceae—Cornaceae, Michigan Flora. Cranbrook Institute of Science and University of Michigan Herbarium, Ann Arbor, Michigan, USA.

Voss, E.G. 1996. Part III, Dicots, Pyrolaceae—Compositae. Michigan Flora. Cranbrook Institute of Science and University of Michigan Herbarium, Ann Arbor, Michigan, USA.

Wakeley, J. 2002. *Developing a "regionalized" version of the Corps of Engineers wetlands delineation manual: Issues and recommendations.* ERDC/EL TR-02-20. Vicksburg, MS: U.S. Army Research and Development Center.

Wakeley, J., and T. Roberts. 1996. Bird distributions and forest zonation in a bottomland hardwood wetland. *Wetlands* 16:296–308.

Walbridge, M. 1993. Functions and values of forested wetlands in the southern U.S. *Journal of Forestry* 91:15–19.

Walton, R., R. Chapman, and J. Davis. 1996b. Development and application of the wetlands dynamic water budget model. *Wetlands* 16:347–357.

Walton, R., J. Davis, T. Martin, and R. Chapman. 1996a. Hydrology of the Black Swamp wetlands on the Cache River, Arkansas. *Wetlands* 16:279–287.

Wamsley, T.V., M.A. Cialone, J.M. Smith, J.H. Atkinson, and J.D. Rosati. 2010. The potential of wetlands in reducing storm surge. *Ocean Eng* 37(1): 59–68.

Wang, C., M. Menenti, M. Stoll, E. Belluco, and M. Marani. 2007. Mapping mixed vegetation communities in salt marshes using airborne spectral data. *Remote Sensing of Environment* 107:559–570.

Wang, Y. 2010. *Remote sensing of coastal environments.* Boca Raton, FL: CRC Press.

Ward, A., and W. Elliot. 1995. *Environmental hydrology.* Boca Raton, FL: CRC/Lewis Publishers.

Ward, A., and S. Trimble. 2004. *Environmental hydrology.* 2nd ed. Boca Raton, FL: CRC Press.

Wdowinski, S., S. Kim, F. Amelung, T. Dixon, F. Miralles-Wilhelm, and R. Sonenshein. 2008. Space-based detection of wetlands' surface water level changes from L-band SAR interferometry. *Remote Sensing of Environment* 112:681–696.

WDR (World Disasters Report) 2010, International Federation of Red Cross and Red Crescent Societies.

Welch, R., M. Remillard, and J. Alberts. 1992. Integration of GPS, remote sensing, and GIS techniques for coastal resource management. *Photogrammetric Engineering and Remote Sensing* 58:1571–1578.

Whigham, D., D. Weller, A. Jacobs, T. Jordan, and M. Kentula. 2003. Assessing the ecological condition of wetlands at the catchment scale. *Landscape Ecology* 20:99–111.

Wilber, D., R. Tighe, and L. O'Neil. 1996. Associations between changes in agriculture and hydrology in the Cache River Basin, Arkansas, USA. *Wetlands* 16:366–378.

Wilcox, D. 1995. Wetland and aquatic macrophytes as indicators of anthropogenic hydrologic disturbance. *Natural Areas Journal* 15:240-248.

Wilen, B. 1990. *The U.S. Fish and Wildlife Service's National Wetlands Inventory. Federal Coastal Wetland Mapping Programs.* Biology Report 90 (18). Washington, DC: Department of Interior, U.S. Fish and Wildlife Service.

Wilhelm, G., and D. Ladd. 1988. Natural area assessment in the Chicago region. In *Transactions of the fifty-third North American Wildlife and Natural Resources Conference, Louisville, Kentucky.* Washington, DC: Wildlife Management Institute.

Williams, D., and J. Lyon. 1991. Use of a geographical information system data base to measure and evaluate wetland changes in the St. Marys River, Michigan. *Hydrobiologia* 219:83–95.

Williams, D., and J. Lyon. 1997. Historical aerial photographs and a GIS to determine the effects of long-term water levels on wetlands along the St. Marys River, Michigan. *Aquatic Botany* 58:363–378.

Willis, C., and W. Mitsch. 1995. Effects of hydrology and nutrients on seedling emergence and biomass of aquatic macrophytes from natural and artificial seed banks. *Ecological Engineering* 4:65–76.

Wold, S. 1995. PLS for multivariate linear modeling. In H. Van de Waterbeemd, ed., *Chemometric methods in molecular design methods and principles in medicinal chemistry.* Weinheim, Germany: Verlag-Chemie.

Wolter, P., C. Johnston, and G. Niemi. 2005. Mapping submerged aquatic vegetation in the U.S. Great Lakes using QuickBird satellite data. *International Journal of Remote Sensing* 26: 5255–5274.

Wright, C., and A. Gallant. 2007. Improved wetland remote sensing in Yellowstone National Park using classification trees to combine TM imagery and ancillary environmental data. *Remote Sensing of Environment* 107:4582–605.

Wu, S. 1989. Multipolarization P-, L-, and C-band radar for coastal zone mapping: the Louisiana example. 1989 Fall Convention of ACSM/ASPRS, Cleveland, Ohio.

Yang, C., J. Everitt, R. Fletcher, J. Jensen, and P. Mausel. 2009. Mapping black mangrove along the south Texas gulf coast using AISA+ hyperspectral imagery. *Photogrammetric Engineering and Remote Sensing* 75:425–436.

Yang, J., and F. Artigas. 2010. Mapping salt marsh vegetation by integrating hyperspectral and LiDAR remote sensing. In J. Wang, *Remote Sensing of Coastal Environment.* Boca Raton, FL: CRC Press.

Yi, G., D. Risley, M. Koneff, and C. Davis. 1994. Development of Ohio's GIS-based wetlands inventory. *Journal of Soil and Water Conservation* 49:23–28.

Yoder, C. 1991. Answering some concerns about biological criteria based on experiences in Ohio. In *Proceedings of Water Quality Standards for the 21st Century.* Washington, DC: U.S. Environmental Protection Agency.

Young, S., and C. Wang. 2001. Land-cover change analysis of China using global-scale Pathfinder AVHRR Land cover (PAL) data, 1982–92. *International Journal of Remote Sensing* 22:1457–1477.

Zandbergen, P., and F. Petersen. 1995. The role of scientific information in policy and decision-making: The lower Fraser basin in transition. Surrey, British Columbia. Symposium and Workshop, Kwantlen College.

Zar, J. 1984. *Biostatistical analysis*. Englewood Cliffs, N.J.: Prentice Hall.

Zomer, R., A. Trabucco, and S. Ustin. 2009. Building spectral libraries for wetlands land cover classification and hyperspectral remote sensing. *Journal of Environmental Management* 90:2170–2177.

Index

A

Accuracy assessment
  of categorized images, 92–95
  in invasive plant mapping, 105–107
  of land cover map, 70, 73–75
  lessons learned, 78–80
Acid mine drainage aerial photography, 20
Activity detection, 169–171; *see also* Change detection methods
ADAR, *see* Airborne Data Acquisition and Registration (ADAR™) System 5500
Advanced Very High Resolution Radiometer (AVHRR), 18, 122
  image preprocessing, 122–123
  resolution of, 161
  sensors, 163
Aerial imaging
  for periodic monitoring, 41
  sensors, 164
Aerial photography
  historical image use, 36–40
  monitoring by, 18
  remote sensing product, 17
  sources of historical images, 37
  stereo, 63–64
  wetlands identification from, 40
Aerial Photography Field Office as source of historical aerial images, 37
Agreement monitoring by remote sensing, 17
Agricultural development effect on wetland condition, 187
Agricultural Stabilization and Conservation Service as source of historical aerial images, 38
Air photo sensors, 164
Air pollution monitoring methods, 18
Airborne Data Acquisition and Registration (ADAR™) System 5500, 98

assessing results with, 107–108
Airborne Observation Platform sensors, 19
Airborne Radar sensors, 165
Airborne sensor imagery remote sensing product, 17
Airborne spectral sensors, 82
Amphibian diversity indicator, 138
Anaerobic condition effects, 5
Anas platyrhynchos, see Mallard duck
Animal habitat, *see* Habitat
Anthropogenic stressors, 127
Applicability of landscape metrics *vs.* measurability, 89
Aqua NASA satellites, 118
Aquatic plant habitat, 183
Aquifer recharging, 24
ArcGIS functions, 73
Archival photographs as data source, 93
ArcInfo software, 83, 87
ArcView 9, 67
ArcView software, 83, 87, 193
Arkansas specific data sources, 179
  habitat database, 181
Assessment of wetlands; *see also* National Wetlands Inventory (NWI)
  National Wetland Condition Assessment (NWCA), 21
  regional, 21
ASTER sensors, 125, 162, 163
Atmospheric correction, 65–66
Atmospheric effects on change detection methods, 121
Atmospheric haze reduction, 123
ATtILA software, 193
AVHRR, *see* Advanced Very High Resolution Radiometer (AVHRR)
Avian dispersal of plant guilds, 48, 52, 54
AWiFS

Soil characterization by remote sensing, 58–59

Soil loss as critical threat, 26

Soil maps, 162, 167

Soil Survey Geographic Database (SSURGO), 167

Solar factors in image interpretation, 121

SOLEC, *see* State of the Lakes Ecosystem Conference (SOLEC)

Southeastern Texas isolated wetlands case study, 60–80

Space Shuttle topography data, 124

Spaceborne advanced sensors, 125

Spatial data sets
  for planning purposes, 171, 173
  resources necessary for, 128

Spatial database needs, 17

Spatial filtering of map information, 87

Spatial miss-registration, 121

Species habitat information sources, 189

Spectral bandpass differences complicate interpretation, 120

Spectral characteristics for plant differentiation, 105

Spectral clustering algorithm, 91

Spectral variation correction, 123

SPOT, 93
  sensors, 162, 163

SRTM, *see* Shuttle Radar Topographic Mapping (SRTM) NASA mission

SSURGO, *see* Soil Survey Geographic Database (SSURGO)

Stakeholder concept, 27

Stand, definition, 100

State of the Lakes Ecosystem Conference (SOLEC), 89

Statistical procedures for mapping data, 87–88

STATSGO, *see* U.S. General Soil Map (STATSGO)

Statview software, 193

Stereo aerial photography, 63–64

Storm indicator, 139

Stressor-variables in case study, 42

Sturgeon habitat, 183

Submerged aquatic vegetation (SAV) in image interpretation, 35

Submersed aquatic plants in near infrared images, 58

Surrogate variables
  for endpoint determination, 9
  use of, 17

Survey sampling techniques where empirical methods not feasible, 16

Surveying data tied to location in photointerpretation, 36

Sustainability principles, 26–27

Sustainability terminology, 25

Sustainable land management (SLM) approaches, 26–27

Synthetic aperture RADAR (SAR)
  elevation data, 162
  sensors, 165
  usefulness of, 161

T

Target plant species, definition, 100

Taxa richness values, 48, 52

Technique integration, 2–3

Temporal change evaluation, 115–119

Temporal data collection, 85

Temporary wetlands, *see* Ephemeral wetlands

Terra NASA satellites, 118

Texture in image interpretation, 32

Thematic mapping, 60, 61
  categories of information, 87
  structure of maps, 87

Thermal infrared sensor use, 161

Time-dependent data collection, 85

Timescale of wetland change, 158–159

TOMS, *see* Total Ozone Mapping Spectrometer (TOMS)

Tone in image interpretation, 32

Topographic data from LIDAR, 125

Total nitrate indicator, 139

Total Ozone Mapping Spectrometer (TOMS), 19

Training data categorization, 90

Training set development, 91

Transboundary approach, 29

Transitional ecological zone
  components of, 11
  techniques inappropriate for, 89

*For Product Safety Concerns and Information please contact
our EU representative GPSR@taylorandfrancis.com Taylor & Francis
Verlag GmbH, Kaufingerstraße 24, 80331 München, Germany*

T - #0151 - 230425 - C81 - 234/156/15 - PB - 9781138076099 - Gloss Lamination